The Hybrid Seller

How AI and Human Behavior Are Redefining Modern Selling

By

Manuel Ferrer, Ed.D.

The Hybrid Seller™
How AI and Human Behavior Are Redefining Modern Selling

Printed in the United States of America

First Edition

ISBN 979-8-9985217-3-7 paperback
ISBN 979-8-9985217-4-4 hardcover
ISBN 979-8-9985217-5-1 ebook
Library of Congress Control Number: 2026902641

TABLE OF CONTENTS

DEDICATION

For every seller doing the hard things the right way.

A NOTE TO THE READER

This book was born out of a paradox I have watched play out in organization after organization, year after year.

We have more data than any generation of sellers in history. More dashboards. More intent signals. More AI-generated insights. And yet, for all of that, revenue stays stubbornly flat. Deals stall without explanation. Conversations that felt like breakthroughs quietly disappear. The pipeline looks full on paper and hollow in reality.

I started asking a different question. What if the problem was never the data?

I learned the answer, in a way, long before I ever carried a quota. I watched my grandfather sell. He didn't run sequences or track engagement metrics. He didn't have a CRM or a tech stack or an AI assistant to prioritize his accounts. What he had was something harder to replicate and easier to underestimate: he paid attention. He listened not just for what people said but for what they meant. He adjusted. He showed up with intention, day after day, in ways that people remembered. He sold through presence.

Years later, when I went deep into applied behavioral science, I found the framework to describe what my grandfather was doing intuitively. Small, repeatable shifts in behavior can permanently change outcomes. That principle holds in therapy, in leadership, and in sales. The mechanisms are the same. The stakes are just different.

The gap isn't in the data. It's in the doing.

This book is for frontline sellers (SDRs, account executives, enterprise reps) who operate in complex, hybrid environments where buyers research independently, engage sporadically, and decide without warning. It is also for the sales leaders, managers, and enablement professionals who are responsible for improving those outcomes but frustrated by the distance between strategy and execution.

If you care less about activity metrics and more about what sellers actually do when it matters, you will find a common language here for coaching, alignment, and sustained performance.

Everything in these pages (the H.Y.B.R.I.D. Framework, the Insight-Execution Gap, and Behavioral Consistency) exists for one purpose: to help you show up with clarity, consistency, and trust in a hybrid, AI-enabled world. Not louder. Not busier. Just more deliberately, reliably human.

You already have the tools. What you need now is something else. A way to show up not just with insight, but with clarity under pressure. A way to move from reactive effort to repeatable influence. A way to stop chasing more and start executing better.

Let's close the gap between knowing and doing—one behavior at a time.

Welcome to The Hybrid Seller.

Introduction
Behavior, Not Bandwidth

It took me longer than I expected to admit that something wasn't working. From the outside, everything looked right. The pipeline was full, activity levels were strong, and the team was doing what they had been trained to do. We had better tools than we had ever had before, more data, more visibility, and more ways to understand what was happening inside the business. The dashboards were cleaner, the reporting was sharper, and the conversations about performance felt more informed than they had in years. If you had looked at it from a distance, you would have assumed we were operating at a higher level than ever before.

And yet, when you stayed close to the work long enough, a different pattern began to emerge.

At first, it showed up in ways that were easy to dismiss. A deal that looked strong would stall without a clear reason. A conversation that felt like it had real momentum would quietly go inactive. A seller would walk out of a meeting confident about what came next, only to find themselves circling back days later without a meaningful response. None of these moments were dramatic on their own, and that was part of what made them difficult to confront. They didn't point to a single breakdown or a clear failure. They accumulated, subtle and persistent, until they began to shape the overall performance of the system.

What became difficult to ignore was not the presence of these moments, but the pattern they created. The outcomes weren't unstable in a way that triggered alarm. They were inconsistent in a way that resisted explanation.

One deal would move forward with ease while another, with similar conditions, would lose momentum. One conversation would create clarity and alignment while another would leave the buyer uncertain, even when the same seller was involved. The more we tried to isolate the cause, the less precise the answers became. It was possible to describe what had happened after the fact, but much harder to understand why it had happened in the moment.

Like most teams in that position, we did what seemed logical. We assumed the issue was visibility. If we could see more, track more, and understand more about what buyers were doing, we could reduce the inconsistency and bring more control to the process. It was a reasonable assumption, and one that aligned with the direction the industry was already moving. So we leaned further into the tools. We expanded our dashboards, layered in additional data, introduced new signals designed to capture buyer interaction, and built reporting structures that offered a more detailed view of the pipeline than we had ever had before.

For a period of time, this felt like progress. The language of our conversations improved, our reviews became more structured, and we had more information to support the decisions we were making. It gave us the sense that we were getting closer to understanding what was actually driving outcomes. But as that initial confidence settled, a different reality began to take hold. The increase in information did not produce the clarity we expected. It introduced a different kind of friction, one that was less visible but more persistent. The more we could see, the harder it became to determine what actually mattered. Signals began to compete with each other, activity blurred into noise, and instead of simplifying decisions, the added visibility made them heavier.

That shift revealed itself in a way that was both simple and frustrating. It became increasingly difficult to answer a basic question: what should we do next?

What became clear in that moment was that the issue wasn't a lack of insight. We had more clarity than ever. The problem was something else.

We weren't consistently translating that clarity into action when it mattered. Not in the dashboards. Not in the planning. But in the moment where a decision had to be made.

That pattern shows up everywhere once you start looking for it.
Sellers learn what works. They intend to apply it. But under pressure, behavior drifts back to what's familiar.

This is what I call the **Insight–Execution Gap**.

Not a knowledge problem. A behavioral one.

It wasn't that we lacked options. If anything, we had too many. The challenge was determining which action would actually move a deal forward and which would simply create the appearance of progress. That distinction, which should have become clearer with more information, became harder to make in the moments that mattered most. And it was in those moments, not in the dashboards or the reports, where performance was ultimately decided.

That realization forced a different line of thinking. The issue was no longer whether we had enough insight. It was whether insight was the problem we were supposed to be solving in the first place.

PART ONE

The New Era of Sales

Mindset, Behavior & Frameworks

Chapter 1:
The Hybrid Modern Seller

The Extinction of One-Channel Selling

Romeo had stared at that dashboard for more than a decade.

Dozens of charts. Color-coded forecasts. Activity counts. Pipeline aging reports. It all looked the way it was supposed to look, the kind of thing you could screenshot for a quarterly review and no one would ask hard questions. It gave the appearance of control, which for a long time had been enough.

But it never answered the only question that actually mattered.

What should I do next?

There was a time when that question didn't exist, at least not in the same way. He flew in, sat across from someone, paid attention to how they responded, and adjusted in real time. The work wasn't simple, but it was direct. He could feel when a conversation was moving forward and when it wasn't. He didn't need a system to interpret it for him.

Now the work had shifted.

His most important relationships lived on a screen. Conversations stretched across channels. Signals appeared outside of meetings and often before they even began. The instincts he had spent twenty years refining were still there, but they had fewer places to land.

So he did what most experienced sellers do when the environment changes.

He looked for clarity in the data.

And the more he looked, the harder it became to find.

Across town, DJ refreshed his dashboard and watched intent signals update in real time.

Page views. Webinar replays. Social activity. AI-ranked accounts. He had access to more information than any seller in the history of the profession. The system was designed to surface opportunity, to make it easier to know where to focus.

At first, it felt like an advantage.

Then it started to feel like pressure.

Every signal carried a sense of urgency. Every notification suggested that something required action. The more he tried to respond to all of it, the more fragmented his day became. He moved quickly, but not always with direction. Activity stayed high, but progress felt uneven.

Because data, on its own, doesn't tell you what to do.

It tells you what happened.

What happens next is still a matter of judgment.

And judgment, without a way to apply it consistently, begins to feel like guesswork.

Somewhere between them stood Fiona.

She was responsible for the number, but more than that, she was responsible for the system that produced it. She could coach metrics. She could walk through a pipeline and identify where deals were at risk. She could talk through strategy and provide direction that made sense in a structured conversation.

What she couldn't see clearly was what was happening inside the interaction itself.

The moment when a buyer hesitated and the response that followed. The timing of a follow-up that either maintained momentum or allowed it to fade. The difference between a question that opened something up and one that kept the conversation at the surface.

Those were the moments that shaped the outcome.

They were also the hardest to observe.

Without that visibility, coaching stayed at the level of outcomes. She could explain *what* needed to happen. She couldn't always show *how* it should happen when it mattered.

Three different roles. Three different strengths.

One shared frustration.

The game had changed faster than they had.

Selling used to follow a pattern that was easy to recognize.

You prospected. You pitched. You presented. You closed. The buyer depended on you for information, and that dependency gave the seller leverage. You controlled the flow of the conversation because you controlled access to what the buyer needed to know.

And then that dependency disappeared.

Today's buyers move through most of their decision process without you. By the time someone appears on your calendar, their thinking is already in motion. They have done the research, compared alternatives, and had internal conversations that would have once included you. What you experience as the beginning of a deal is often much closer to the middle.

This changes the role of the seller in ways that are easy to underestimate.

You are no longer guiding from the start.

You are entering a process already shaped by inputs you did not control.

It would be easy to assume that more information solves this problem.

That if you can see what the buyer is doing, track their activity, and analyze their behavior, you can regain control of the process. That assumption has driven most of the investment in sales technology over the past decade.

More data. More signals. More visibility.

And to a point, it works.

It tells you where activity is happening. It shows you patterns you might otherwise miss. It creates the sense that the system is becoming more knowable.

But it doesn't answer the question Romeo asked.

What should I do next?

Because information does not create action.

Behavior does.

There is a moment in most sales careers when you realize that the seller who wins is not always the one with the best pitch or the deepest product knowledge. It is the one who follows up when others forget, who personalizes when others rely on templates, who slows down when the conversation requires it instead of rushing to the next step.

These are not dramatic differences.

They are small behaviors.

What makes them powerful is not their size, but their consistency.

From a behavioral standpoint, this is where performance begins to take shape.

In Applied Behavior Analysis, outcomes are not treated as isolated events. They are the result of patterns that repeat over time. A behavior occurs in response to a situation, produces a result, and, if that result is reinforcing, becomes more likely to occur again. With repetition, those actions form patterns, and those patterns produce outcomes.

The same structure exists in selling.

A signal appears. A buyer hesitates, engages, asks a question, or goes quiet. The seller responds. That response either creates clarity or introduces friction. As those responses repeat the deal begins to take shape around them.

When those behaviors are inconsistent, outcomes feel unpredictable.

When they stabilize, outcomes begin to follow.

This is where the Hybrid Seller begins to separate.

Not in the tools they use or the channels they operate in, but in how they behave within them.

They recognize that signals matter, but they do not react to them automatically. They interpret them within context. They understand that a page view is not intent, that a delayed response is not necessarily disengagement, and that not every signal requires action.

Instead, they apply a simple filter, often without realizing it.

What changed?

Why does it matter now?

What is the right action in this moment?

The answer is rarely complex.

A well-timed message. A specific observation. A question that surfaces something the buyer hasn't yet said.

Nothing dramatic.

But when those behaviors repeat, they create something the market consistently rewards.

Trust.

Romeo began to see this in his own work.

The difference between deals that moved forward and those that stalled was not the strategy. It was the consistency of behavior within it. In the deals that progressed, there was alignment. He responded in a way that matched what the moment required. In the deals that didn't, the behaviors were still there, but applied at the wrong time or with less precision.

The difference was subtle.

The impact was not.

This is where the conversation around AI needs to be reframed.

AI is not the advantage.

It does not replace judgment. It does not build trust. It does not navigate the complexity of a real conversation.

What it does, when used well, is make behavior visible.

It shows patterns that would otherwise go unnoticed. It highlights how often certain actions occur, how consistently they are applied, and what tends to follow them. It closes the gap between what a seller believes they are doing and what is actually happening.

And once behavior becomes visible, it can be shaped.

This is the foundation of **Behavioral Consistency**.

Not a new methodology, but a different way of understanding performance.

The focus shifts from what the seller knows to what the seller does, repeatedly, across time and across situations. It shifts from isolated moments of execution to patterns of behavior that hold under pressure.

From a productivity standpoint, this changes everything.

More activity does not necessarily produce better outcomes.

More consistent behavior does.

The Hybrid Seller operates with that understanding.

The difference rarely comes down to knowing more. It comes down to what you do in the moment. The small, consistent micro-behaviors that shape how a conversation unfolds.

They do not attempt to control the environment. They focus on how they respond within it. They recognize the moments that matter and bring attention to those moments consistently. They do not rely on a single strong interaction. They build patterns that repeat.

As those patterns take hold, variability begins to narrow.

They do not eliminate uncertainty, but they make performance more stable.

Looking back, what Romeo was searching for was not a better dashboard.

It was a better way to navigate the work.

He didn't need more information.

He needed a clearer way to act on what he already had, and a way to do it consistently enough that the outcome was not left to chance.

That is where this book begins.

Not with a new process, but with a shift in perspective.

Because once behavior becomes visible, and once it is applied with consistency, the question of what to do next becomes easier to answer.

Not because the environment is simpler.

But because the response to it is more stable.

Behavioral Recap: What to Practice This Week

This week, pay attention to the moments where you default to what is familiar instead of what the situation actually calls for. That is where the real work is.

- Catch yourself when you reach for the same channel or the same follow-up approach out of habit. Notice it before you send it.

- When a buyer goes quiet, pause before you push. Silence is not always rejection. It is often processing.

- Pick one channel you rarely use this week and reach out through it intentionally. Not because the platform matters, but because the behavior of adapting does.

- Before each interaction, ask yourself where this buyer actually is right now, not where your pipeline says they should be.

- After each conversation, ask one honest question: did that move the deal forward, or did it just fill time?

These are not dramatic changes. They are the small behavioral shifts that, practiced consistently, separate sellers who produce reliably from those who produce in bursts.

> **Quick Self-Check:** *This week, how often did I adapt my behavior to what the moment required, instead of defaulting to what felt comfortable?*

Next Chapter: The Buyer's Journey Has Changed—Have You?

Chapter 1 showed you why the old playbooks stopped working. Not because sellers forgot how to sell, but because buyers stopped moving the way sellers expected.

Chapter 2 takes you inside the buyer's world. The loops, the pauses, the silent research, the internal debates happening long before anyone picks up the phone. You will learn why great meetings still lead to stalled deals, how momentum disappears without a single objection, and where sellers misread buyers and pay for it later.

The Hybrid Seller does not push buyers forward. They recognize where buyers already are and adjust accordingly. That shift, from driving to reading, is what Chapter 2 teaches.

Chapter 2:
The Buyer's Journey Has Changed— Have You?

How Sellers Lose Momentum Without Ever Realizing Why

Romeo didn't realize he had lost the deal until it was already over.

The meeting had gone well. At least by every standard he had learned to trust. The conversation was fluid, the buyer was engaged, and the questions that surfaced felt like the kind that typically led somewhere. There were no obvious objections, no tension in the room, nothing that suggested resistance. When the call ended, he did what he had always done. He documented the notes, set a follow-up, and moved on to the next priority with the quiet confidence that this one was progressing the way it should.

Three days passed without a response. That wasn't unusual. A week passed, and the follow-up he sent felt slightly heavier than the first. By the second week, the tone had shifted. It was no longer a continuation of momentum, but an attempt to reestablish it.

The reply, when it came, was brief and polite. They had decided to move in a different direction.

There was no explanation beyond that. No objection to address, no feedback to refine. Just a closed door that, from his vantage point, had never appeared to be closing.

What made it more difficult to process was what he saw a few days later. The same buyer, or at least someone from the same organization, shared an update publicly. They were excited to be moving forward with a solution they had been evaluating for months.

Months.

That was the part that stayed with him.

Not the loss itself, but the realization that the decision had been forming long before he ever entered the conversation. What he had experienced as the beginning of a deal was, in reality, much closer to the end of one.

This is the part of modern selling that takes time to accept.

The buyer's journey no longer begins when you make contact. In many cases, it is already well underway. Buyers research independently, compare options, seek input from peers, and form early conclusions before they ever speak to a seller. By the time a meeting is scheduled, a significant portion of the decision-making process has already taken place.

What makes this difficult is not just the loss of control, but the loss of visibility. Much of what influences the decision happens outside of the interaction. It happens in internal conversations, in side discussions, in moments of uncertainty that are rarely expressed directly. From the seller's perspective, it can feel as though the deal is moving in one direction, when in fact it has already shifted in another.

This is often described as a *nonlinear journey*, but that description, while accurate, does not fully capture what is happening. The movement is not simply nonlinear. It is layered with competing inputs, shifting priorities, and a level of cognitive load that most sellers underestimate.

Buyers are not moving through a clean sequence of stages. They are navigating a series of loops.

They revisit information. They reconsider earlier assumptions. They seek validation, encounter conflicting perspectives, and adjust their thinking in response. At times, progress accelerates. At others, it slows or stops entirely. From the outside, it can appear unpredictable. From the inside, it reflects the complexity of making a decision that carries risk.

There is a tendency to interpret this behavior as indecision or lack of urgency. In reality, it is often the opposite.

The modern buyer operates in an environment saturated with information. Each input requires interpretation. Each option introduces trade-offs. Each decision carries implications that extend beyond the immediate purchase. In that context, hesitation is not a sign of disengagement. It is a sign of processing.

This is where the behavioral lens becomes useful.

In Applied Behavior Analysis, behavior is understood in relation to the environment in which it occurs. Actions are shaped by stimuli, influenced by competing demands, and reinforced by outcomes that are not always immediately visible. When applied to buying behavior, this perspective shifts the focus away from isolated actions and toward the conditions that produce them.

A delayed response is not simply a lack of interest. It may reflect competing priorities, internal alignment challenges, or uncertainty that has not yet been resolved. A request for more information is not always a request for detail. It can be an attempt to reduce perceived risk. Even silence, which is often interpreted as disengagement, can signal a pause in the decision process rather than its end.

When these behaviors are viewed in isolation, they are easy to misinterpret. When viewed as part of a pattern, they begin to make more sense.

The challenge for the seller is not simply to recognize these patterns, but to align behavior with them.

This is where many deals begin to drift.

The seller operates from a linear model. The buyer operates from a looping one. The seller attempts to advance the conversation. The buyer is still processing earlier information. The result is a misalignment that is rarely acknowledged directly, but felt in the interaction.

It often shows up as a subtle shift in tone. The conversation becomes less fluid. Responses become shorter. Follow-ups require more effort. From the seller's perspective, it can feel as though the deal is losing momentum without a clear reason. From the buyer's perspective, it can feel as though the conversation is moving faster than their ability to make sense of it.

That gap is not created by a lack of effort. It is created by a lack of alignment between behavior and context.

Romeo began to see this only after he changed the way he reviewed his own deals.

Instead of focusing on outcomes, he started to look at sequences. What happened before the conversation, during it, and after it. He paid closer attention to the moments that had once felt insignificant. A pause in a response. A question that didn't fully land. A follow-up that arrived later than it should have.

What he found was not a single point of failure, but a pattern.

In the deals that moved forward, there was a consistent alignment between his behavior and the buyer's state. He slowed down when the conversation required space. He asked questions that helped the buyer clarify their own thinking. He followed up in ways that reinforced momentum rather than attempting to create it artificially.

In the deals that stalled, that alignment was less consistent. The behaviors were not dramatically different, but they were applied at the wrong time or with less precision. The same follow-up that worked in one context felt premature in another. The same level of detail that was helpful for one buyer created overload for another.

The difference was not in the strategy.

It was in the timing and consistency of behavior.

This is where the connection to productivity becomes more explicit.

In a traditional model, productivity is often measured through activity. More calls, more emails, more meetings. The assumption is that increased activity leads to increased opportunity, which in turn leads to improved outcomes.

But when behavior is misaligned with the buyer's context, increased activity can produce diminishing returns. More outreach does not necessarily create more response. More meetings do not necessarily create more progress. In some cases, they create additional friction by introducing information faster than the buyer can process it.

From a behavioral standpoint, productivity is better understood as the efficiency of action relative to outcome. It is not the volume of behavior, but the precision and consistency of it. A well-timed question that clarifies uncertainty can move a deal forward more effectively than a series of follow-ups that attempt to force progress.

This does not reduce the importance of effort. It refines how effort is applied.

Technology, and particularly AI, plays a role here, but not in the way it is often described.

The value is not in accelerating every action. It is in helping identify where action is most effective. By surfacing patterns in engagement, response timing, and progression, it provides a clearer picture of how buyers are actually moving. It allows sellers to see where behavior is aligned and where it is not.

In behavioral terms, it enhances the ability to observe the relationship between stimulus and response. It makes it easier to identify which actions are reinforcing progress and which are not. This creates the conditions for more consistent behavior.

But the technology does not make the decision.

It informs it.

The responsibility for alignment still rests with the seller.

The shift, then, is not simply in understanding that the buyer's journey has changed.

It is in recognizing that selling must change in response.

Not by adding more steps or more structure, but by developing the ability to read the situation more accurately and respond with greater precision. By recognizing that movement in a deal is not driven by force, but by clarity. And that clarity is often created through behaviors that are small enough to be overlooked, but significant enough to shape the outcome.

By the time Romeo began to adjust in this way, the experience of a stalled deal started to feel different.

It was no longer something that happened to him.

It was something he could examine.

He could trace it back to specific moments, specific behaviors, and specific misalignments. And in doing so, he gained something he had not had before.

Not certainty.

But control over the one variable that mattered most.

How he showed up when it counted.

Behavioral Recap: What to Practice This Week

The work this week is about alignment, not acceleration. Before you advance a deal, check whether the buyer is actually ready to move.

- Look at each active deal and identify where the buyer genuinely is right now. Not where your CRM stage says they are. Where their behavior suggests they are.

- Match your outreach depth to buyer readiness. A buyer exploring needs something different than a buyer deciding. Treat them differently.

- Translate your next steps into the buyer's language, not your sales process language. If your next step sounds like a stage gate, rewrite it.

- Verify intent before assuming it. Politeness is not progress. Engagement is not commitment. Know the difference.

These behaviors will not always feel productive in the moment. They will feel like slowing down. But slowing down to read the buyer accurately is what makes everything afterward faster.

Quick Self-Check: *How accurately did my actions this week reflect where the buyer actually was, rather than where I wanted them to be?*

Next Chapter: The H.Y.B.R.I.D. Framework

Chapter 2 gave you a clearer picture of how buyers actually move. Chapter 3 gives you the structure to respond to that movement, consistently and deliberately.

The H.Y.B.R.I.D. Framework is the strategic map Hybrid Sellers use to interpret buyer signals in real time, choose the right behavior for the moment, stay human without becoming inconsistent, and move deals forward without pressure.

Understanding the buyer is the starting point. Chapter 3 is where that understanding becomes actionable.

Chapter 3:
The H.Y.B.R.I.D. Framework
From Chaos to Clarity

Romeo didn't set out to build a framework. If anything, he would have resisted the idea. He had spent enough time in sales to know how most frameworks were received. They sounded clear in a training room, made sense on a slide, and then quietly disappeared once the work began. The problem was never the logic behind them. It was the distance between the model and the moment when it needed to be applied.

What he was trying to solve felt more immediate than that.

He wanted a way to make sense of what he was seeing in real time. Not after a deal was lost or won, but while it was still unfolding. He wanted to understand why certain conversations moved forward and others stalled, even when the variables appeared similar. And more than that, he wanted a way to respond that didn't rely on instinct alone.

The turning point came during a deal that, by any traditional measure, should have worked.

It was complex, high visibility, and involved multiple stakeholders. The kind of opportunity that would have once been managed through a series of in-person sessions, where alignment could be built gradually and reinforced through presence. Instead, every interaction took place across screens. Conversations were shorter. Participation varied. Decision-makers appeared and disappeared without warning. The structure that had once supported the process was no longer there.

At one point, midway through the cycle, Romeo paused and said something that felt both obvious and unsettling.

"I don't know where we actually are."

It wasn't a statement about the stage of the deal. It was something more specific than that. He didn't know how the buyer was thinking. He didn't know what had been resolved and what was still uncertain. He didn't know which conversations mattered most or which ones were simply maintaining the appearance of progress.

In the absence of that clarity, it became difficult to decide how to act.

And that is where most deals begin to drift.

What followed was not a reinvention of the sales process, but a gradual shift in how he approached each interaction. Instead of trying to force the deal into a predefined structure, he began to pay closer attention to where the buyer actually was and what the moment required. The ability to read and adjust in real time is not a single skill. It's a series of small, observable micro-behaviors that signal awareness and control. That meant adjusting pacing, changing the way he framed questions, and rethinking how he used each channel. It also meant letting go of the assumption that progress had to look a certain way.

As these behaviors repeated, a pattern began to emerge.

Certain behaviors consistently created clarity. Others introduced friction, even when they were well intentioned. Some actions helped the buyer move forward. Others caused them to pause, not because the solution was wrong, but because the interaction had added complexity at the wrong time.

What made the pattern useful was not that it was new, but that it was repeatable.

He began to see that the difference between movement and stagnation was not tied to a single tactic, but to how a set of behaviors aligned with the buyer's context across the life of the deal. When that alignment was present, progress felt natural. When it was not, no amount of effort could compensate for it.

That realization became the foundation of what would later be described as the H.Y.B.R.I.D. Framework.

It is important to understand what this framework is *not*.

It is not a script, and it is not a sequence of steps to follow. It does not replace judgment, nor does it attempt to simplify a process that is inherently complex. Instead, it provides a way of organizing attention. It offers a set of reference points that help the seller determine where to focus and how to respond, moment by moment, as a deal unfolds.

Each element reflects an aspect of behavior that, when applied consistently, contributes to clarity and forward movement.

The first of these is **Human Connection**.

In a traditional sense, this might be described as rapport or relationship building. In practice, it is something more specific. It is the accumulation of behaviors that signal reliability. The consistency of follow-through. The ability to recall and reference prior conversations. The adjustment of tone and pacing in response to the situation. Over time, these behaviors create a sense that the seller is present, attentive, and aligned.

From a behavioral standpoint, this can be understood as a form of *reinforcement*. When a buyer experiences consistency in these interactions, it reduces uncertainty. That reduction in uncertainty makes it easier to continue engaging, which in turn strengthens the relationship. The connection is not built in a single moment. It is shaped through repeated, predictable behavior.

The second element is an understanding of the **Buyer's Journey**, not as a static model, but as something interpreted through signals.

By this point, it should be clear that the buyer's movement is not linear. The challenge is not recognizing that, but responding to it accurately. Signals appear continuously, but they do not carry meaning on their own. It is the interpretation of those signals that determines the next action.

A spike in engagement may suggest interest, or it may reflect internal comparison. A delayed response may indicate hesitation, or it may be unrelated to the decision entirely. The seller's role is not to react to the signal, but to interpret it within context and choose a behavior that reduces uncertainty.

In behavioral terms, this is the distinction between *stimulus* and *response*. The signal is the stimulus. The seller's action is the response. As consistency holds, the effectiveness of that response can be observed and adjusted. The more accurately the response aligns with the situation, the more likely it is to be reinforced through progress.

The third element is Bond Building.

While similar to **Human Connection**, it operates at a different level. If connection is about reliability, bonding is about memory. It is the set of behaviors that make the interaction distinct enough to be remembered. A thoughtful follow-up that references a specific concern. A brief message that demonstrates understanding of the buyer's environment. A moment where the seller steps outside of the expected pattern to provide something of value.

These behaviors function as reinforcers in a different way. They create a positive association that extends beyond the immediate interaction. Gradually, that association influences how future interactions are perceived. The seller is no longer just a participant in the process, but a source of clarity and support.

The fourth element is **Relevance**.

In an environment saturated with information, relevance is not defined by personalization alone. It is defined by timing and context. A well-crafted message delivered at the wrong moment can create friction. A simple observation delivered at the right moment can create momentum.

Relevance requires an understanding of what the buyer is trying to resolve and how that objective is evolving. It also requires restraint. Not every insight needs to be shared. Not every capability needs to be presented. The behavior that creates value is the one that aligns with the buyer's immediate need, not the seller's agenda.

From a behavioral perspective, relevance can be seen as *matching*. The closer the behavior aligns with the context, the more likely it is to be reinforced. When the alignment is off, even slightly, the interaction loses effectiveness.

The fifth element is **Influence**.

In many discussions, influence is treated as a function of persuasion. Here, it is better understood as the ability to create movement. A buyer does not move forward because they have received more information. They move forward because something has become clearer. That clarity reduces the perceived risk of action.

The behaviors that support this are often subtle. Framing a decision in a way that highlights trade-offs. Asking a question that surfaces an unspoken concern. Offering a perspective that helps the buyer make sense of competing inputs. These actions do not push the deal forward directly. They create the conditions in which forward movement becomes possible.

In behavioral terms, this can be linked to *shaping*. Each interaction nudges the buyer's understanding in a direction that makes the next step easier to take.

The final element is **Driving Value**.

This is often discussed in the context of post-sale activity, but its impact begins much earlier. Value is not something that is delivered at the end of the process. It is something that is demonstrated throughout it. Each interaction either adds clarity or adds noise. With time, those interactions shape the buyer's perception of the relationship.

When value is delivered consistently, it reinforces trust. It creates a sense that the seller is contributing to the outcome, not simply facilitating it. That perception carries forward, influencing not only the current deal, but future opportunities as well.

Taken together, these elements do not form a sequence.

They form a map.

They provide a way of orienting behavior within a complex environment. They help the seller determine where to focus, what to observe, and how to respond. They do not eliminate uncertainty, but they reduce it. And in doing so, they make consistency more achievable.

What makes this framework effective is not the ideas themselves, but the way they are applied.

A single instance of any one of these behaviors is unlikely to have a significant impact. Their value emerges through repetition. When they are applied consistently, across different deals and different contexts, patterns begin to stabilize. The seller becomes more predictable in the way they respond, and that predictability creates trust.

This is where the connection to productivity becomes clear.

When behavior is aligned and consistent, less effort is required to achieve the same outcome. Conversations become more efficient. Decisions are made with greater confidence. Momentum is maintained with less intervention. Productivity improves, not because more is being done, but because what is being done is more effective.

Looking back, Romeo realized that what he had been searching for was not a better way to manage the pipeline, but a better way to navigate it.

The structure he had built was not something he followed step by step. It was something he returned to when the path was unclear. It gave him a way to interpret what he was seeing and to choose his response with greater intention.

As a result, that was enough to change the pattern.

Not by eliminating uncertainty, but by providing a consistent way to move through it.

Behavioral Recap: What to Practice This Week

This week, use the H.Y.B.R.I.D. Framework as a diagnostic tool, not a checklist. Pick one deal that has stalled and run it through each pillar.

H—Have you created the kind of reliability this buyer would recognize and remember, or have you been efficient but forgettable?

Y—Do you actually know where this buyer is in their decision process right now, or are you operating on an assumption?

B—Have you done something recently that was distinctly personal and unexpected, something that made this interaction stand out from every other vendor conversation they are having?

R—Is your last message genuinely relevant to what this buyer is dealing with today, or is it relevant to what you want to sell them?

I—Are you creating forward movement and clarity, or are you maintaining presence without producing progress?

D—Have you delivered something of value in the last thirty days that had nothing to do with closing the deal?

Whichever pillar produces the weakest honest answer is where your behavior needs to go this week.

Quick Self-Check: *When I look at my active deals, which H.Y.B.R.I.D. pillar is most consistently absent, and what would it take to bring it in?*

Next Chapter: Behavioral Consistency

You now have the map. Chapter 4 builds the engine that makes consistent navigation possible.

Behavioral Consistency is what turns the H.Y.B.R.I.D. Framework from a concept you understand into a set of behaviors you execute reliably, even under pressure, even on a hard week, even when the deal is not moving and the instinct is to just do more.

Chapter 4 will show you how top performers are not more gifted. They are more repeatable. And repeatability is something you can build.

Chapter 4:
Behavioral Consistency
Turning High-Impact Behaviors into Repeatable Results

By the time Fiona began to change the way she coached, she had already come to an uncomfortable conclusion. The issue wasn't effort, and it wasn't talent. If anything, that made the problem harder to accept. She had people on her team who could run a great call, handle a difficult conversation, and build trust quickly when everything aligned. She had seen it happen often enough to know the capability was there. What she couldn't rely on was when it would show up.

One week, a seller would execute at a high level, asking the right questions, following up with clarity, moving a deal forward with a kind of quiet precision that made the process feel almost simple. The next week, in a similar situation, that same seller would miss something small. A follow-up would come later than it should have. A conversation would stay on the surface instead of going deeper. A next step would be implied instead of confirmed.

None of it was dramatic. That was the problem.

It was just enough to change the outcome.

And because those differences were subtle, they were easy to explain away. Timing. The buyer. Internal factors outside of anyone's control. All of those explanations were valid, at least in part. But they didn't fully account for what Fiona was seeing.

The variation wasn't random.

It was patterned.

Sellers knew what to do. They had demonstrated it before. But they weren't doing it consistently when it mattered.

This is where most performance breaks down. Not in knowing, but in doing. It's the Insight–Execution Gap in practice.

She started to notice this gap in the way deals moved. In some cases, progress felt steady. Not fast, not forced, but consistent. Each interaction built on the last. Questions led to clarity. Clarity led to decisions. Even when obstacles appeared, they were addressed in a way that kept the process moving.

In other cases, the pattern was different. The deal would start with energy, then lose momentum in small increments. A delay here. A missed detail there. Nothing that would stand out in isolation, but enough to create friction. Little by little, that friction accumulated, and what had once felt like progress became something closer to drift.

The difference between those two patterns was not the strategy being used.

It was the consistency of behavior within it.

In behavioral science, there is a principle that becomes difficult to ignore once you see it in action. Behavior is not defined by what someone is capable of doing at their best. It is defined by what they do repeatedly, across time and under varying conditions.

This distinction matters more in sales than most people realize.

A single well-executed interaction can create the impression of skill. A series of consistently executed interactions creates performance. The gap between the two is where most of the variability in outcomes originates.

When behavior is inconsistent, outcomes will be inconsistent, even if the underlying knowledge and ability remain constant. When behavior stabilizes, outcomes begin to stabilize with it.

That is the foundation of Behavioral Consistency.

Fiona's shift was simple in concept, but difficult in practice.

She stopped asking what happened in the deal and started asking what the seller did, specifically and in sequence. Not in general terms, but in detail. When did the follow-up go out? How was the next step framed? What question was asked when the buyer hesitated? What changed in the interaction after that point?

At first, the answers were vague.

"I followed up quickly."

"I handled the concern."

"We aligned on next steps."

Those responses sounded right, but they didn't provide anything that could be observed or improved. So she kept pressing.

"How quickly?"

"What exactly did you say?"

"What did the buyer do after that?"

It was in those details that the patterns began to appear.

Once behavior became visible, something important happened.

It became measurable.

Not in the sense of assigning a score to every interaction, but in the sense of identifying whether a behavior was present, how often it occurred, and what tended to follow it. Repeatedly, certain **micro-behaviors**, the small, observable actions within an interaction, began to stand out. Not because they were complex, but because they appeared consistently in the deals that moved forward.

Follow-up latency was one of them. When sellers responded within a predictable window, momentum was easier to maintain. When that timing slipped, even slightly, the interaction lost continuity.

The **depth of discovery** was another. When sellers moved beyond the surface and helped the buyer articulate what was actually driving the decision, the conversation gained clarity. When they stayed at a higher level, uncertainty remained.

The way **next steps** were handled made a similar difference. When they were stated clearly and confirmed, the process had direction. When they were implied, the burden shifted back to the buyer, and progress slowed.

These weren't the only ones, but they were part of a small set of repeatable micro-behaviors that consistently separated momentum from drift.

Individually, none of these micro-behaviors would be described as advanced. They weren't new. And they weren't difficult.

What made them impactful was their consistency.

This is where the connection to productivity becomes more direct.

In many sales environments, productivity is framed as a function of activity. More calls, more outreach, more meetings. The assumption is that increased activity increases the likelihood of success. There is some truth to that, particularly at the early stages of the pipeline.

But as deals progress, the relationship changes.

At that point, productivity is less about volume and more about precision. The number of actions matters less than the quality and timing of those actions. A single well-executed follow-up can maintain momentum more effectively than multiple attempts that arrive too late or lack clarity.

From a behavioral standpoint, productivity improves when high-impact behaviors are executed consistently, not when overall activity increases. The focus shifts from doing more to doing what matters, reliably.

The challenge, of course, is that consistency is difficult to maintain.

Under pressure, behavior tends to revert to habit. When time is limited, when multiple deals are competing for attention, when outcomes feel uncertain, people default to what is easiest, not necessarily what is most effective.

This is not a failure of discipline. It is a characteristic of behavior.

In Applied Behavior Analysis, this is often described in terms of reinforcement. Behaviors that are followed by a positive outcome are more likely to be repeated. Behaviors that are not reinforced, or are inconsistently reinforced, are less stable. As a result, this creates patterns that can either support performance or undermine it.

In a sales context, the reinforcement is not always immediate or obvious. A well-timed question may contribute to a decision that occurs weeks later. A delayed follow-up may weaken momentum without creating a clear point of failure. This makes it difficult to connect behavior to outcome without a structured way of observing the relationship.

This is where a simple loop becomes useful.

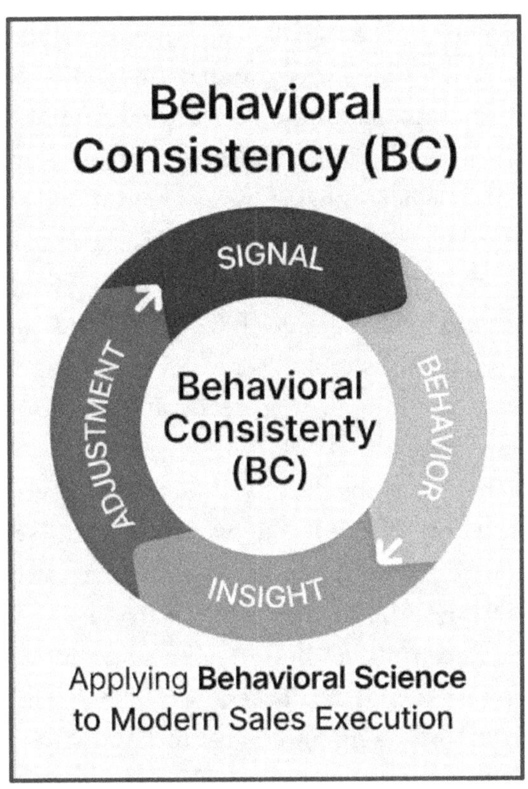

Figure 4.1 — Behavior Consistency Loop

A signal appears, the seller responds with a behavior, that behavior produces an outcome, and over time patterns emerge that inform the next action. Most sellers operate within the first two steps. They notice the signal and respond. Fewer examine what follows, and fewer still use that information to adjust behavior deliberately. The result is that the same patterns repeat, often without awareness.

The breakdown, however, does not occur in theory. It occurs in execution, in the moment where the seller must act. It happens mid-call, when the buyer hesitates, when the conversation shifts, and when the pressure to respond increases. This is where behavior begins to drift, and where top performers make small adjustments that change the outcome. It is also where Behavioral Consistency is most difficult to maintain, not after the call and not in a review, but in real time, while the outcome is still being shaped.

This is where most systems begin to fall short.

They work well when there is time to think. They support planning, review, and reflection. They help sellers understand what should have happened and, in many cases, what they should do next.

But the moment that determines the outcome rarely happens there.

It happens inside the interaction itself, where the conversation shifts without warning and the seller is forced to respond in real time. A buyer hesitates, asks a question that reframes the discussion, or introduces a concern that had not been visible before. The seller recognizes it, but recognition alone is not enough. The response has to match the moment, and it has to happen quickly enough to matter.

This is where most sellers default. Not because they lack knowledge, but because the conditions do not allow for careful analysis. The pressure to respond compresses decision-making, and behavior tends to revert to habit.

It is also where Behavioral Consistency is most difficult to maintain.

Not after the call.

Not in a dashboard.

But in the moment where the deal is still being shaped.

Fiona began to see a shift in how her team operated. Not immediately, and not uniformly, but gradually. Conversations about performance became more specific. Feedback moved from general advice to observable actions. Sellers began to recognize their own patterns, not just in isolated moments, but across deals. What changed was not the level of effort. It was the reliability of execution. And as that reliability improved, so did the outcomes, not in dramatic spikes, but in a steadier pattern that was easier to sustain.

Looking back, the change was less about introducing something new and more about bringing attention to something that had always been there. Behavior had always driven performance. What was missing was a way to see it clearly enough to shape it, and a way to adjust it while the moment was still unfolding, when it still had the chance to change.

This is the gap **Agent Insight™** was built to address.

Agent Insight is the real-time behavioral companion app to The Hybrid Seller, designed to sit alongside the seller during active conversations and provide immediate, behavior-focused guidance. It is not a system for storing information or tracking activity, and it does not attempt to replace the seller's judgment. Its role is more specific than that. It exists to help the seller recognize what is happening in the moment, interpret it accurately, and adjust before the opportunity is lost.

In practice, this means narrowing the focus to a single question: *what is the most effective behavior right now?*

Agent Insight does not overwhelm you with options. It sharpens your attention. It translates signals into action, helping you move from awareness to execution without delay.

Agent Insight organizes that decision across six execution domains:

- Alignment Check (Are we solving the right problem?)
- Discovery Depth (Do I fully understand their situation?)
- Value Clarity (Can they clearly see the impact?)
- Deal Momentum (Is this actually moving forward?)

- Follow-Up Discipline (Have I defined the next step?)
- Deal Control (Am I leading the process?)

These domains are not theoretical. They map directly to the moments where deals either progress or stall. Each one acts as a lens, helping the seller diagnose what is happening beneath the surface of the conversation.

Within each domain are targeted prompts and behavioral adjustments designed to guide action in real time. Not scripts to memorize, but decision pathways that help the seller respond with precision.

You are not searching for answers.

You are tightening the moment.

Eventually, something subtle begins to happen. The pause between signal and response becomes more deliberate. Reactions become decisions. Behaviors that were once inconsistent begin to stabilize.

This is the shift.

From knowing to executing under pressure. From post-call feedback to in-the-moment adjustment. From activity to consistent behavior.

Agent Insight does not replace judgment.

It sharpens it.

Scan this to access Agent Insight before your next call.

If the loop explains how behavior improves in the long run, Agent Insight exists to shape it in the moment.

Behavioral Recap: What to Practice This Week

Pick one behavior this week and practice it with full intention. Not a list of improvements. One.

- Choose the micro-behavior that, if you executed it reliably every time, would most directly affect your results. Follow-up timing. Discovery depth. Next-step confirmation. Pick the one that matters most right now.

- Track execution, not outcomes. Did the behavior happen? Yes or no. The results will follow, but start by measuring the action.

- After each interaction, note what the buyer did immediately following your behavior. Eventually, this teaches you what actually creates momentum in your specific market.

- At the end of the week, make one small adjustment based on what you observed. Not a reinvention. A refinement.

This is how consistency is built. One behavior, tracked honestly, adjusted deliberately. The compounding effect is real, but it requires patience with the process.

Quick Self-Check: *This week, how disciplined was I about practicing consistency rather than chasing intensity?*

Next Chapter: Consultative Selling in the Hybrid Era

Behavioral Consistency tells you how to show up. Chapter 5 shows you how to show up in the conversation itself.

The hybrid era did not kill selling. It killed the pitch. Buyers today are informed, skeptical, and running out of patience for sellers who talk at them rather than with them. Chapter 5 is where you learn to sell like a guide instead of a guru, to ask questions that clarify rather than impress, and to replace persuasion with progress.

Because in the hybrid era, trust is not built by being the smartest voice in the room. It is built by being the most useful one, at the moment it matters most.

PART TWO
Core Selling Skills & Technology
Human First, Enhanced by AI

Chapter 5:
Consultative Selling in the Hybrid Era
When Diagnosis Becomes Your Differentiator

Romeo used to believe that a strong presentation could carry a deal.

It wasn't an unreasonable belief. For a long time, it had worked. If you understood your product, anticipated objections, and delivered a clear, structured narrative, you could move a conversation forward. Buyers came to meetings expecting to learn, and the seller's role was to guide that learning.

At least, that's how it used to feel.

The shift was gradual enough that it was easy to miss. Buyers started showing up differently. They asked more specific questions, referenced competitors without prompting, and moved through the early parts of the conversation with a level of familiarity that would have been unusual before. What had once been an introduction became something closer to a confirmation.

It's in moments like this that small adjustments begin to matter more, and where tools like Agent Insight create separation, helping sellers recognize what is happening and respond before the conversation slips.

Romeo adjusted the way most experienced sellers do. He refined the presentation, tightened the language, anticipated the new questions. The mechanics improved. The delivery became more efficient.

But the outcomes didn't follow in the way he expected.

There were still meetings that felt strong in the moment and led nowhere afterward. There were still conversations that seemed aligned but failed to produce a clear next step. The difference was that the gap between what he delivered and what the buyer needed had become harder to detect.

It took a while to see the pattern clearly.

The issue wasn't that the presentation was weak.

It was that the presentation was no longer the point.

The realization didn't come during a major deal or a high-pressure moment. It came during a call that, on the surface, looked routine. The buyer had already done their research. They understood the general value proposition, were familiar with the product, and had a clear sense of what they were trying to evaluate.

Romeo moved through the early part of the conversation the way he always had, setting context, confirming understanding, and guiding the discussion toward the areas he believed mattered most. The buyer listened, asked a few questions, and responded in a way that suggested alignment.

But something felt off.

The conversation was moving, but not deepening. Information was being exchanged, but it wasn't creating clarity. The interaction had the structure of progress without the substance of it.

At one point, Romeo paused and asked a question he might have overlooked in the past.

"What's making this decision difficult right now?"

The buyer didn't answer immediately. There was a brief silence, followed by a response that shifted the entire direction of the conversation.

"It's not the solution," they said. "It's whether we can actually implement it without creating more problems."

That was the moment the conversation became useful.

What had been missing up to that point was not information. It was interpretation.

The buyer did not need more detail about the product. They needed help making sense of their own situation. The hesitation they were experiencing was not about features or pricing. It was about risk, and more specifically, the risk of disruption.

Once that was surfaced, the conversation changed. It moved away from describing what the solution could do and toward understanding how it would fit within the buyer's environment. The focus shifted from presentation to diagnosis.

This is the core of consultative selling, but in the hybrid environment, it takes on a different level of importance.

Buyers are no longer looking for information in the same way they once did. They can access that on their own, often before the conversation even begins. What they lack is clarity. They are navigating a set of inputs that are often conflicting, incomplete, or difficult to reconcile. In that context, the seller's role changes.

It becomes less about delivering content and more about helping the buyer think.

From a behavioral standpoint, this shift is significant.

The traditional model reinforces behaviors associated with presentation. Delivering information clearly, addressing objections directly, moving efficiently toward a close. These behaviors are still relevant, but they are no longer sufficient on their own. The behaviors that now carry more weight are those that uncover, clarify, and guide.

Listening becomes more than a passive activity. It becomes a way of identifying what has not been said.

Questions are no longer just a means of gathering information. They become tools for shaping the buyer's understanding. Even silence, when used deliberately, can create space for the buyer to articulate something they had not fully considered.

These behaviors are subtle, but they are measurable in their effect.

When a seller listens at a deeper level, the buyer's responses change. They become more specific, more reflective, and more aligned with the underlying issue. When questions are framed in a way that encourages exploration rather than confirmation, the conversation gains depth. When the seller resists the urge to fill every gap with information, the buyer has room to process.

These interactions begin to produce a different pattern.

The conversation becomes more efficient, not because it is faster, but because it is more focused. Decisions are made with greater confidence, not because more information is provided, but because the right information is clarified.

Romeo began to notice that the most effective conversations were not the ones where he spoke the most, but the ones where he created the conditions for the buyer to think more clearly.

That required a shift in behavior.

He had to slow down in moments where his instinct was to move forward. He had to ask one more question when the surface answer felt sufficient. He had to resist the urge to demonstrate value through information and instead allow it to emerge through understanding.

These are not dramatic changes. They do not require a new skill set as much as a refinement of an existing one.

But they are not automatic.

They require consistency.

This is where the connection to productivity becomes clearer again.

In a traditional model, productivity might be measured by how efficiently a seller can move through a presentation or how many conversations they can manage in a given period of time. In a consultative model, productivity is better understood as the ability to move a conversation toward clarity.

A single question that reframes the buyer's thinking can create more progress than a series of statements that reinforce what they already know. A well-timed pause can surface a concern that would otherwise remain hidden. These moments are not accidental. They are the result of behaviors that are applied deliberately and consistently.

From a behavioral perspective, these actions function as *interventions*. They alter the trajectory of the interaction in a way that makes progress more likely. When those interventions are applied with enough consistency, they begin to shape the overall pattern of the deal.

Technology can support this process, but again, its role is limited.

AI can provide context, surface relevant information, and even suggest potential questions based on prior interactions. What it cannot do is determine which question matters in the moment or how it should be delivered. That requires judgment, and judgment is built through experience and reinforced through behavior.

What AI can do, when used effectively, is help identify which behaviors are associated with successful outcomes. It can show patterns across conversations, highlight where depth was achieved, and where it was not. This information can then be used to refine the seller's approach, making it more consistent over time.

Fiona noticed this shift as she observed her team.

The sellers who began to adopt a more consultative approach did not necessarily have more conversations, but the conversations they had were different. They produced clearer next steps, more engaged buyers, and a stronger sense of alignment. As those differences accumulated, a pattern emerged.

What changed was not the volume of activity.

It was the quality and consistency of behavior within each interaction.

Looking back, the distinction becomes straightforward.

In the hybrid era, the value of the seller is not determined by how much they can tell the buyer.

It is determined by how effectively they can help the buyer understand.

And that understanding is created through behavior.

Not once, but repeatedly, across every conversation that matters.

Behavioral Recap: What to Practice This Week

The Guide Check: Ask This Mid-Call Before You Move On

Most sellers assess a call after it ends. By then, the moment has passed. The Hybrid Seller makes a brief internal check while the conversation is still unfolding.

When you sense a conversation starting to drift, when responses are getting shorter or the buyer sounds like they are waiting for you to finish, pause and ask yourself two questions silently:

1. Is what I am saying helping this buyer think more clearly about their actual situation?

2. Is this moving the decision forward in a way that is meaningful to them, not just to me?

If the answer to either question is no, do not accelerate. Instead:

- *Slow the pace and create a beat of silence*
- *Ask one deeper question, not a clarifying one, a genuinely curious one*
- *Restate what you heard before adding anything new*
- *Make the next step smaller and easier to say yes to*

AI call coaching tools can flag moments where talk time spikes or buyer engagement drops. Use those cues to identify where the Guide Check was needed, and train yourself to catch it earlier next time.

Next Chapter: Mastering Omnichannel Sales

Consultative selling shapes how you behave in a single conversation. Chapter 6 expands the frame to the full landscape of interactions across which that conversation now lives.

Buyers do not experience selling as a sequence of separate touchpoints. They experience it as one continuous impression. Chapter 6 shows how Hybrid Sellers create coherence across channels, align timing and message to where the buyer actually is, and build momentum without overwhelming attention.

This is where individual conversations become a selling rhythm. And where behavioral consistency turns omnichannel complexity into competitive advantage.

Chapter 6:
Mastering Omnichannel Sales

Meeting Buyers Where They Are—and Where They Actually Are

The first time DJ realized something had changed, it wasn't during a call.

It was in between them.

He had always thought of selling as a series of conversations. Calls, meetings, presentations, follow-ups. Each interaction had a clear beginning and end, and if you managed those moments well, the deal moved forward. What happened outside of them mattered, but it felt secondary, something that supported the main event.

That distinction had started to fade.

He noticed it in small ways at first. A buyer referencing something he had posted online. Another mentioning a piece of content they had seen weeks before the first meeting. A third who had clearly formed an opinion before the conversation even began, based on interactions DJ hadn't directly participated in.

None of it was part of his formal process.

And yet, it was shaping the outcome.

The shift is easy to describe once you see it, but difficult to account for in real time.

Selling no longer happens in a single channel, or even in a clearly defined sequence of channels. It unfolds across a series of touchpoints that are loosely connected but collectively influential. A message here, a post there, a follow-up that arrives at the right moment, an interaction that reinforces something the buyer is already considering.

From the buyer's perspective, it is experienced as a continuous flow.

From the seller's perspective, it can feel fragmented.

That fragmentation is where consistency begins to break down.

DJ had always been disciplined about his outreach. He followed the structure he had been taught, moved prospects through sequences, and ensured that no opportunity went untouched. But as the number of channels increased, so did the number of decisions he had to make.

Should he send an email or a message through another platform?

Should he follow up immediately or wait for a better moment?

Should he reference prior interactions or introduce something new?

Each decision was small, but together they shaped the way the buyer experienced the interaction.

At first, DJ approached this the way most sellers do.

He tried to optimize each channel independently.

He improved his email language, refined his messages, experimented with content. In isolation, each improvement made sense. But the overall experience remained uneven. A strong message in one channel could be followed by a generic follow-up in another. A thoughtful interaction could be diluted by a sequence that felt automated.

The issue was not the quality of any single action.

It was the lack of consistency across them.

From a behavioral perspective, this is where context becomes critical.

Behavior does not exist in isolation. It is influenced by the environment in which it occurs, and in a multi-channel setting, that environment is constantly shifting. Each channel carries its own expectations, its own pace, and its own form of feedback. What feels appropriate in one may feel out of place in another.

When behavior is not aligned across those contexts, it creates a disjointed experience.

For the buyer, this shows up as inconsistency. The seller feels attentive in one moment and distant in another. The message feels tailored in one interaction and generic in the next. Even when each action is reasonable on its own, the overall pattern lacks coherence.

That lack of coherence introduces friction.

The sellers who adapt most effectively to this environment tend to make a different shift.

Instead of thinking in terms of channels, they think in terms of continuity.

Each interaction is not treated as a separate event, but as part of an ongoing conversation. The choice of channel becomes secondary to the quality and consistency of the behavior being expressed through it.

A follow-up is not just a task to complete, but a continuation of something that has already been established. A message is not an isolated outreach, but a reinforcement of context. A piece of content is not simply shared, but positioned in a way that aligns with what the buyer is already considering.

This does not require a dramatic increase in effort.

It requires a different orientation.

DJ began to see this when he looked at his own interactions more closely.

In the deals that progressed, there was a sense of continuity. The tone, timing, and content of his communication aligned across channels. Each touchpoint reinforced the last. The buyer did not need to reorient themselves with each interaction, because the thread of the conversation remained intact.

In the deals that stalled, that continuity was less consistent. The connection between interactions weakened. The buyer was required to do more work to make sense of the conversation, and gradually, that effort reduced their willingness to continue.

The difference was not in the number of touchpoints.

It was in how they connected.

This is where Behavioral Consistency extends beyond the individual interaction and into the broader system.

It is not enough to execute a behavior well once. It must be executed in a way that aligns with what came before and supports what comes next. The pattern that forms across those interactions is what shapes the buyer's experience.

From a productivity standpoint, this has important implications.

When behavior is consistent across channels, fewer touchpoints are needed to maintain momentum. Each interaction builds on the last, reducing the need to reestablish context. The conversation progresses with less effort because it is not being reset repeatedly.

When behavior is inconsistent, more effort is required to achieve the same result. Additional outreach is needed to compensate for lost momentum. Messages are repeated, clarified, and reinforced, not because they were ineffective, but because they were not connected.

In this way, inconsistency creates inefficiency, even when activity levels remain high.

Technology plays a supporting role here as well.

AI can help identify which channels are being used, how often, and with what level of interaction. It can surface patterns in response rates and highlight where interactions are gaining or losing traction. But again, it does not determine how those channels should be used.

What it can do is reveal whether behavior is consistent across them.

It can show whether follow-ups maintain the same level of relevance, whether timing aligns with prior buyer signals, and whether the overall pattern supports continuity. That information can then be used to adjust behavior in a way that brings the system into alignment.

Fiona saw this shift across her team.

The sellers who improved were not necessarily the ones who adopted every new channel or experimented with every available tool. They were the ones who brought consistency to the channels they were already using. They ensured that each interaction, regardless of where it occurred, reflected the same level of attention and alignment.

That consistency became visible in their results.

Conversations required less rework. Buyers stayed engaged for longer periods. Deals moved with a steadier pace, not because they were being pushed, but because they were being supported.

Looking back, DJ realized that what had changed was not the number of ways he could reach a buyer, but the importance of how those interactions fit together.

The work had become less about managing channels and more about managing behavior within them.

And once that shift took hold, the complexity of the environment became easier to navigate.

Not because there were fewer decisions to make, but because those decisions were guided by a consistent approach.

Behavioral Recap: What to Practice This Week

This week, shift your focus from channel activity to channel continuity. The goal is not to be in more places. It is to make every touchpoint feel like part of the same conversation.

- Design your outreach across at least two channels for your top five active accounts. Think about how each touchpoint connects to the last, not just what it says on its own.

- Match the channel to what the moment actually requires. Use email when the buyer needs clarity in writing. Use video when nuance matters more than efficiency. Use social presence for warmth and visibility.

- Reduce volume and raise relevance. Before you send a follow-up, ask: why this message, why this channel, why now, for this specific person?

- After you send a touchpoint, pause before the next one. Give the buyer space to respond rather than stacking messages that signal pressure.

- Make sure every follow-up answers one question: *what does this buyer need to hear from me at this specific moment in their decision process?*

Next Chapter: Leveraging AI, Data, and Automation with Heart

Chapter 6 showed you how to create coherence across channels. Chapter 7 examines the tools that increasingly shape how sellers show up across all of them.

AI and automation are not going away. The question is whether you use them in a way that strengthens your behavior or one that quietly replaces it. Chapter 7 draws that line clearly. You will learn how to use AI to surface signals rather than shortcuts, translate data into better behavioral decisions, automate support without automating sincerity, and design systems that reinforce your best habits instead of eroding your judgment.

Done well, AI does not make selling colder. It makes the best sellers more deliberate.

Chapter 7:
Leveraging AI, Data, and Automation with Heart

AI, Automation, and the Discipline of Consistency

Fiona didn't set out to bring more technology into the team.

If anything, she was cautious about it. By the time she began to rethink how her team was operating, they already had access to more tools than they could consistently use. Each one promised a different advantage. Better targeting. Smarter sequencing. More efficient follow-up. On their own, none of those promises were unreasonable. Taken together, they created something harder to manage.

The problem was not capability.

It was application.

She had seen what happened when a new tool was introduced. There was a period of enthusiasm, followed by a period of experimentation, and then a gradual return to familiar patterns. The tool remained, but the behavior it was meant to support did not always follow. As a result, the system became layered with functionality that existed in theory more than in practice.

So when the conversation turned to AI, her first instinct was to be careful.

Not because she doubted its potential, but because she understood where the breakdown usually occurred. It wasn't in what the tool could do. It was in how consistently it was used to support behavior that actually mattered.

The early use cases followed a familiar path.

AI was introduced to generate outreach, summarize conversations, and recommend next steps. It reduced the time required to complete certain tasks and provided a level of support that would have been difficult to replicate manually. For sellers managing multiple deals, this created an immediate sense of relief. The workload felt lighter, and the process more manageable.

But as Fiona observed how the team was using it, a pattern began to emerge.

The sellers who relied on it to replace thinking saw limited improvement. Their output increased, but the quality of their interactions did not change in a meaningful way. Messages were sent more quickly, but they did not create more responses. Follow-ups were more frequent, but they did not consistently move deals forward.

In contrast, the sellers who used AI differently began to show a shift.

They treated it as a way to sharpen their behavior rather than substitute for it. They used it to identify where they might be missing something, to refine how they framed a question, or to ensure that their follow-up aligned with what had actually been said. The tool did not change what they were trying to do. It helped them do it with greater precision.

The difference between these two approaches became increasingly apparent.

From a behavioral perspective, this distinction is important.

Tools can increase the frequency of a behavior, but they do not determine its effectiveness. If the underlying behavior is misaligned, increasing its frequency can amplify the problem rather than solve it. Sending more messages that lack relevance does not create more interaction. It creates more noise.

For behavior to produce a consistent outcome, it must be both appropriate to the context and reinforced over time. AI can support the first by providing information and context. It can support the second by making patterns visible. What it cannot do is replace the need for the behavior itself.

This is where the role of AI becomes clearer.

It is not the driver of performance.

It is the amplifier of behavior.

When used well, it reduces the effort required to execute high-impact micro-behaviors consistently. It helps ensure that follow-ups are timely, that messages reflect the context of the conversation, and that important details are not overlooked. It brings a level of structure to the process that makes consistency easier to maintain.

But it does not remove the need for judgment.

In fact, it increases it.

Because the presence of more information requires a clearer sense of what to act on and what to ignore.

It's in this space, where judgment becomes more important, that tools like Agent Insight reinforce the discipline of behavior, helping sellers focus on what matters without outsourcing the decision itself.

Fiona began to frame this differently with her team.

Instead of introducing AI as a way to do more, she positioned it as a way to do fewer things better. The goal was not to increase output, but to improve the consistency of the behaviors that mattered most. That meant being selective about where the tool was applied and intentional about how it supported the interaction.

For example, instead of generating a complete message, a seller might use AI to refine a single sentence that clarified the next step. Instead of relying on a summary, they might use it to identify a moment in the conversation that warranted a deeper question. In each case, the tool supported the behavior, but did not replace it.

This approach required more discipline than automation.

It also produced better results.

The connection to productivity becomes more apparent in this context.

When AI is used to increase activity, productivity gains are often short-lived. The initial efficiency is offset by the need to manage a higher volume of interactions that may not produce meaningful progress. This can create a cycle where more effort is required to maintain the same level of performance.

When AI is used to support Behavioral Consistency, the effect is different.

High-impact behaviors become easier to execute reliably. Follow-ups occur within the appropriate window. Messages align more closely with the buyer's context. Conversations maintain continuity across interactions. The overall system becomes more stable, not because more is being done, but because what is being done is more consistent.

This is where the compounding effect begins.

In behavioral science, consistency is often linked to reinforcement schedules. When a behavior is reinforced consistently, it becomes more likely to occur again. When reinforcement is intermittent or unclear, the behavior becomes less stable. In a sales environment, reinforcement is not always immediate, which makes consistency more difficult to maintain.

AI can help bridge that gap by providing feedback that is closer to the behavior itself. It can highlight when a follow-up was delayed, when a message lacked alignment, or when a conversation did not reach the necessary depth. This feedback, when used correctly, serves as a form of reinforcement. It brings attention to the behavior and creates an opportunity for adjustment.

With consistent reinforcement, this process strengthens the behavior.

Fiona noticed that the sellers who improved the most were not the ones who used AI the most aggressively.

They were the ones who used it most deliberately.

They identified a small number of behaviors that had a clear impact on their deals and used the tool to support those behaviors consistently. They did not try to optimize everything at once. They focused on what mattered and applied the same level of attention across interactions.

This created a pattern that was both visible and repeatable.

Their conversations became more aligned. Their follow-ups more timely. Their deals more predictable. Not in the sense that every outcome was known in advance, but in the sense that the variability decreased.

Looking back, Fiona realized that the introduction of AI had not changed the nature of the work.

It had clarified it.

It made it easier to see that performance was not a function of how much could be automated, but how well behavior could be executed consistently within a complex environment.

The technology did not create that discipline. It revealed the need for it.

Behavioral Recap: What to Practice This Week

This week, audit how you are actually using AI and automation. Not whether you are using it, but whether it is making your behavior sharper or replacing it.

- For every piece of AI-generated outreach you send this week, add one sentence that the AI could not have written because it requires your specific knowledge of this buyer.

- When you receive an intent signal or an alert, run it through the three-question filter before you act: *what changed, why does it matter now, what is the right behavior within 48 hours?*

- Use AI to draft the structure of follow-ups. Then read each one aloud before you send it. If it does not sound like you, revise it until it does.

- Vary the type of outreach you use across the week. An insight one day, a question the next, a story or a proof point after that. Variation signals presence. Monotony signals automation.

- When a buyer engages, slow down rather than speeding up. Engagement is an invitation to have a conversation, not a trigger to accelerate the sequence.

Quick Self-Check: *This week, how often did I earn responses by being relevant rather than merely persistent?*

Next Chapter: Hybrid Prospecting

Now that you have the tools and the framework to operate across the full selling landscape, Chapter 8 returns to where every relationship begins: the first contact.

Hybrid prospecting is not cold blasts or spray-and-pray sequences. It is the discipline of showing up at the right moment with the right message, using AI and data to spot real openings and relying on behavior to earn attention and trust. You will learn how to identify the signals that actually matter, open conversations without forcing them, and turn first touches into forward motion rather than resistance.

Prospecting is not about getting louder. It is about getting more precise.

PART THREE

The Hybrid Seller in Action

From First Contact to Closed Deal

Chapter 8:
Hybrid Prospecting
A New Approach to Finding and Engaging Leads

For all the tools, data, and systems that define modern selling, prospecting still comes down to a simple constraint. You are asking for someone's attention.

What has changed is not the nature of that request, but the conditions in which it is made. Attention is no longer limited by access. It is limited by relevance.

Today's seller operates in an environment where information is abundant. Signals are visible in ways they were not before. A company's priorities, hiring patterns, messaging shifts, and buying behaviors can all be observed with a level of detail that was previously unavailable. AI has accelerated this further, making it possible to gather context, identify patterns, and generate outreach at scale. In theory, this should make prospecting easier. In practice, it has made the difference between effective and ineffective prospecting more pronounced.

Because when everyone has access to the same information, the advantage no longer comes from what is available. It comes from how it is interpreted and how it is applied.

Traditional prospecting was built on activity. The logic was straightforward. More calls, more emails, more touches created more opportunities. This approach worked in an environment where access to buyers was more limited and interruption was more accepted. Sellers could rely on persistence to create engagement, and volume could compensate for inefficiency.

That model begins to break down in an environment where buyers are constantly filtering information. They are informed, selective, and accustomed to ignoring anything that does not immediately connect to their current priorities. In this context, more activity does not necessarily create more opportunity. It often creates more noise.

The modern seller faces a different problem. It is not a lack of information. It is the need to determine what matters within it.

AI plays a role in this, but its role is often misunderstood. It can surface signals, identify patterns across accounts, and generate messaging that appears personalized. It can tell you that a company has expanded into a new market, that a leadership change has occurred, or that interaction with a certain type of content has increased. It can suggest what to say and how to say it.

What it cannot do is determine whether any of it is relevant in the moment.

That requires judgment.

The most effective sellers use AI not to replace prospecting behavior, but to sharpen it. They use it to gather context, to identify potential signals, and to reduce the time required to prepare. But the decision of what to act on, when to act, and how to engage remains theirs.

From a behavioral perspective, prospecting is not primarily a messaging exercise. It is a system of actions and outcomes that either reinforce or weaken future behavior.

Every outreach attempt produces a result. There is a response, or there is not. That outcome is not neutral. It shapes what the seller is likely to do next. A message that generates a response is more likely to be repeated. A message that produces silence is often adjusted, abandoned, or replaced.

What is often missing is the explicit connection between the behavior and the outcome. Sellers track activity. They review results. But they do not always isolate which specific behaviors produced which outcomes. Without that connection, prospecting becomes reactive. Adjustments are made based on assumptions rather than patterns.

When that connection is made clear, prospecting begins to change. It becomes less about trying different approaches and more about refining specific behaviors that consistently produce results.

Romeo experienced this shift gradually. Early in his role, his approach to prospecting was disciplined but conventional. He built lists, followed structured sequences, and executed with consistency. His activity levels were high, and his messaging was informed by the information available to him. He knew the accounts he was targeting, understood their general profile, and could articulate the value of what he was offering.

Despite this, his results were uneven.

Some outreach efforts generated responses. Others did not. At first, he attributed this to external factors. Timing, market conditions, the variability of buyer interest. These explanations were not entirely wrong, but they did not fully account for the pattern he was seeing.

The turning point came when he began to look more closely at the interactions that did produce engagement.

What he noticed was not dramatic. The messages were not radically different in structure or length. What changed was the starting point. Instead of reaching out because a contact fit a defined profile, he began reaching out because something had changed.

A company had announced a new initiative. A department was hiring for roles that suggested a shift in priorities. A leader had taken on a new position and was beginning to shape their agenda. These were not abstract data points. They were signals that indicated movement.

He began to anchor his outreach in those signals.

Rather than introducing himself with a general statement of value, he referenced the change directly. He framed his message in a way that reflected what was happening within the organization and connected it to a potential outcome. The goal was not to impress, but to demonstrate relevance.

The response was different.

Not universally, and not immediately, but consistently enough to be noticeable. Conversations began more easily. Replies were more thoughtful. Engagement extended beyond the initial interaction.

From a behavioral standpoint, this created a clear pattern. The signal-informed outreach produced a different outcome than the generic, profile-based approach. That outcome reinforced the behavior that led to it.

As he continued, the pattern became more defined. He became more selective about when to reach out and more precise in how he did it. The volume of his activity did not necessarily increase. In some cases, it decreased. What changed was the quality of the interactions that followed.

This shift did not eliminate the need for effort. It changed where that effort was applied.

Instead of investing time in sending more messages, he invested time in understanding which signals were worth acting on and how to interpret them. AI supported this by making those signals easier to identify, but it did not remove the need for judgment. If anything, it increased the importance of it.

Because more signals do not automatically create more opportunity. They create more decisions.

The discipline required in this environment is not obvious. It involves slowing down when speed feels more productive. It involves choosing not to act on every available signal, but to focus on the ones that are most relevant. It involves accepting that fewer, more aligned interactions can produce better outcomes than a higher volume of disconnected ones.

Over repeated interactions, these choices begin to shape performance.

When behavior is aligned with context, variability decreases. Outcomes become more predictable. Results become more consistent. The process requires less correction, less follow-up to recover lost momentum, and less effort to reestablish relevance.

When behavior is misaligned, the opposite occurs. Activity increases, but results remain inconsistent. Additional effort is required to compensate for interactions that did not connect in the first place.

The difference between these two states is not always visible at the level of individual actions. It becomes clear when patterns are observed over time.

The modern seller operates within that pattern.

They interpret what is happening, choose the right micro-behavior in the moment, and apply it consistently. That consistency reduces variability and creates more stable performance. It is not driven by volume, but by alignment.

Technology will continue to change the way prospecting is executed. New tools will emerge, new signals will become available, and new methods of engagement will be introduced. These changes will alter the surface of the process, but not its foundation.

The ability to recognize what matters, to act in a way that reflects that understanding, and to adjust based on the outcome will remain constant.

Prospecting has always been about creating opportunity. What has changed is how that opportunity is earned.

Behavioral Recap: What to Practice This Week

This week, replace one volume-based prospecting behavior with one signal-based one. Quality of action over quantity of activity.

- Start every outreach sequence with a signal, not a list. What happened at this account recently? What changed? Build from there.

- Lead with context rather than a pitch. Show the buyer that you understand something about their world before you ask anything of them.

- Use permission questions instead of meeting requests on first contact. "Would it be useful if I sent a one-page breakdown?" opens more doors than "Can we jump on a call this week?"

- Send one genuinely human, signal-based touch to your top five high-intent accounts. Not a template. Something that demonstrates you paid attention.

- Match your ask to the buyer's stage and bandwidth. Early-stage buyers need a different level of commitment than buyers who are actively evaluating.

- At the end of the week, review your activity and ask which behaviors produced responses and which ones produced silence. Let the pattern inform next week.

Quick Self-Check: *How often this week did I create forward momentum without relying on pressure or volume?*

Next Chapter: Mastering Hybrid Sales Meetings and Presentations

Prospecting earns attention. Chapter 9 shows you what to do with it.

Hybrid meetings are not demos or deck walkthroughs. They are decisions taking shape in real time, across screens and in rooms, with stakeholders who are half-present, fully skeptical, and more informed than they used to be. Chapter 9 shows you how to lead these meetings with clarity and confidence, how to engage both virtual and in-person participants intentionally, and how to leave every conversation with shared understanding rather than vague next steps.

Great meetings do not persuade. They align. And alignment is what moves deals forward.

Chapter 9:
Mastering Hybrid Sales Meetings and Presentations

Follow-Through and the Preservation of Momentum

If there is a point in the sales process where the outcome becomes visible, it is not in the outreach, and it is not in the follow-up.

It is in the meeting.

Not because the meeting determines the decision on its own, but because it reveals how the decision will be made.

In modern selling, that meeting rarely looks the way it once did.

It is no longer a controlled environment where a seller stands at the front of the room and presents to a group that is fully present. It is a distributed interaction, often across screens, with participants who are partially engaged, managing competing priorities, and evaluating information in real time.

Some are on camera. Some are not. Some are focused. Others are responding to messages, reviewing documents, or joining from environments that make full attention difficult.

The seller is no longer competing with other vendors.

They are competing with everything else happening in the buyer's world at that moment.

This changes the function of the meeting.

It is no longer enough to present clearly.

The modern seller must lead attention, shape interpretation, and create alignment while the interaction is still unfolding.

Romeo recognized this shift.

Early in his role, his meetings followed a familiar pattern. He prepared thoroughly, structured his presentation carefully, and focused on delivering information in a way that was clear and comprehensive. He walked through the material step by step, ensuring that each point was explained and supported.

The meetings were organized. The content was accurate. The delivery was professional.

But the outcomes were inconsistent.

Some conversations progressed. Others stalled. In many cases, the feedback was positive, but the next step was unclear. There was agreement in the moment, but not always commitment afterward.

At first, he attributed this to external factors. The complexity of the decision, the number of stakeholders, the timing of the conversation. These were all part of the environment, but they did not fully explain the pattern.

The shift began when he stopped evaluating the meeting based on how well it was delivered and started evaluating it based on what was happening within it.

He began to pay attention to moments where attention shifted. When cameras turned off. When responses became shorter. When a question was answered, but not fully engaged with. When one participant spoke frequently while others remained silent.

These were not disruptions to the meeting.

They were signals.

From a behavioral perspective, these signals reflect changes in engagement. They indicate whether the interaction is reinforcing participation or allowing it to fade.

The behavior that follows determines whether the meeting moves toward alignment or away from it.

In the past, Romeo would have continued.

He would have moved forward with the presentation, trusting that the structure would carry the conversation. The goal was to complete the material and ensure that everything had been covered.

This time, he adjusted.

When attention shifted, he paused. When a response felt incomplete, he asked a follow-up question. When certain participants remained quiet, he brought them into the conversation directly.

The structure of the meeting did not disappear, but it became secondary to what was happening in real time.

This changed the dynamic.

The meeting became less about delivering information and more about shaping understanding. The pace adjusted. The sequence adapted. The conversation became more responsive to the signals within it.

From the outside, the difference was subtle.

From the buyer's perspective, it was significant.

They were no longer being presented to.

They were being engaged.

This is where the role of technology becomes more precise.

AI and modern tools can enhance meetings, but only when they are used to support the behavior taking place within them.

They can provide real-time insights. They can surface relevant information during the conversation. They can assist with capturing notes, identifying key themes, and ensuring that important points are not lost.

They can also create risk.

A seller who relies too heavily on generated content can become disconnected from the interaction itself. Attention shifts from the conversation to the tool. The response becomes less grounded in what is happening and more influenced by what is suggested.

The most effective sellers use these tools differently.

They prepare with them, but they do not defer to them.

Before the meeting, they use AI to understand the account more deeply. They review recent activity, identify potential areas of focus, and consider how the conversation might unfold. They enter the meeting with context, not just content.

During the meeting, their attention remains on the interaction.

They listen for shifts in tone. They observe changes in engagement. They adjust based on what is said and what is not.

After the meeting, they use technology to capture and reinforce what occurred, but the meeting itself remains a human activity.

This distinction matters.

Because the meeting is not a transfer of information.

It is a sequence of micro-behaviors that either build or weaken alignment.

Fiona began to coach her team with this in mind.

Instead of asking whether the presentation was delivered effectively, she asked what happened within specific moments. Where did attention shift. Who engaged and who did not. What signals were observed, and how did the seller respond.

At first, this level of detail felt unfamiliar.

Sellers are often trained to evaluate meetings in broader terms. Was the buyer interested. Did the conversation go well. Was the next step secured.

These questions are useful, but they do not capture how alignment is created.

By focusing on behavior within the meeting, Fiona helped her team see where alignment was built and where it was lost. They began to recognize patterns. Certain actions consistently produced engagement. Others allowed it to fade.

Over repeated interactions, those patterns became more stable.

The connection to performance becomes clearer in this context.

When a meeting is led effectively, fewer follow-ups are required to clarify what was already discussed. Fewer additional meetings are needed to rebuild understanding. The decision process moves forward with less friction.

When a meeting is not led effectively, the opposite occurs. Information must be repeated. Context must be reestablished. Additional interactions are required to regain alignment.

In this way, the meeting is not a step in the process.

It is the process.

The hybrid seller understands this.

They do not measure success by how much content was covered, but by how clearly the group leaves with a shared understanding of what has been discussed and what will happen next.

They interpret what is happening, choose the right action in the moment, and apply that action consistently.

This reduces variability and creates more stable performance.

Hybrid meetings introduce an additional layer of complexity. Attention is uneven. Participation is distributed. The environment is less controlled.

In these conditions, the seller must be more deliberate.

They must direct questions intentionally. They must create moments for engagement rather than waiting for them to occur. They must ensure that remote participants are included, not as an afterthought, but as a primary part of the interaction.

They must also recognize that silence is not neutral.

It is a signal.

The ability to interpret that signal and respond to it determines whether the meeting continues to move forward or begins to drift.

Technology will continue to shape how meetings are conducted.

Platforms will evolve. Tools will improve. The mechanics of interaction will change.

What will remain constant is the need for the seller to lead.

To create clarity in the moment. To maintain alignment across participants. To ensure that the conversation results in a shared understanding that can be carried forward.

The best meetings do not persuade.

They align.

And alignment, once created, reduces the effort required for everything that follows.

Behavioral Recap: What to Practice This Week

This week, treat every meeting as a leadership moment. Your goal is not to present well. It is to run the room in a way that produces shared clarity and a genuine next step.

- Send a brief pre-meeting note to every call this week confirming the agenda and what you need from the buyer in this conversation. Not a formality. A genuine setup that tells them you take their time seriously.

- In every meeting, ask one question early that reorients the agenda around the buyer's current reality. "Since we last spoke, what has changed for you?" is a version of this.

- After a buyer responds to a question, pause for two seconds before you reply. Give the answer room to land. Give yourself room to actually process what was said rather than transitioning immediately to your next point.

- End every meeting by confirming the next step out loud, specifically, with a name, a date, and an action. Then send a recap within 90 minutes that reflects the buyer's actual words, not your interpretation of them.

- If you are running hybrid meetings with some participants on screen and some in the room, deliberately direct questions and check-ins to the remote participants first. They are the easiest to lose.

Quick Self-Check: *How consistently did I leave my meetings this week with clear shared understanding and an explicit next step, rather than vague positive momentum?*

Next Chapter: Negotiating and Closing Deals in a Hybrid Sales World

You have earned the attention. You have learned to run the room across every format the hybrid world throws at you. Now comes the stage where most deals quietly die: the close.

Chapter 10 shows you how hybrid sellers close without pressure. How to read hesitation before it becomes delay. How to de-risk decisions for buying committees. How to handle objections without flinching and frame value in the buyer's language, not your own. How to turn "we are interested" into a confident, committed yes.

The best closers do not push. They steady the room, align the decision, and make the next step feel safe. Chapter 10 shows you exactly how that is done.

PART FOUR
Beyond the Close
Growth, Leadership, and the Future

Chapter 10:
Negotiating and Closing Deals in a Hybrid Sales World

Decision and the Reduction of Risk

By the time a deal reaches its final stages, most of the visible work has already been done.

The buyer understands the solution. The conversations have covered the major questions. The value has been established, at least in principle. From the seller's perspective, this is often where the process feels most predictable. The path forward appears clear, and the remaining steps seem procedural.

And yet, this is where many deals slow down.

Not because something new has been introduced, but because something unresolved has remained.

Romeo began to recognize this pattern only after losing a deal that, by every logical measure, should have closed.

The buyer had been engaged throughout the process. The interactions were consistent, the concerns were addressed, and the solution aligned with what they were trying to achieve. There had been no indication of resistance, no signals that suggested the deal was at risk.

As the conversation moved toward a final decision, Romeo did what he had always done. He reinforced the value, clarified the terms, and asked for the commitment.

The response was not a rejection.

It was a delay.

The buyer needed more time.

At first, this seemed reasonable. Decisions at that level often required additional consideration. Internal alignment needed to be confirmed, details needed to be finalized, and stakeholders needed to agree. Romeo continued to engage, providing information, answering questions, and maintaining contact.

But the delay did not resolve.

It extended.

What had felt like a temporary pause became something more permanent. The urgency that had been present earlier in the process began to dissipate. The conversations became less frequent, and when they did occur, they carried less momentum.

Eventually, the deal was no longer active.

What made this outcome difficult to understand was that nothing had changed in a visible way.

The solution had not become less relevant. The buyer's needs had not shifted significantly. The conditions that had supported the deal were still in place.

What had changed was the buyer's willingness to act.

This is the point in the process where the nature of the decision becomes more important than the details of the solution.

From the outside, a decision can appear to be a rational evaluation of options. Features are compared, pricing is considered, and a choice is made based on which solution provides the greatest benefit.

In practice, the decision is influenced by something less visible.

Risk.

Not just financial risk, but operational, reputational, and personal risk. The risk of making a decision that introduces new problems. The risk of being responsible for an outcome that does not meet expectations. The risk of change itself.

As the moment of commitment approaches, these considerations become more prominent.

And if they are not addressed directly, they can outweigh the perceived value of the solution.

From a behavioral standpoint, this is a predictable shift.

As the consequence of a decision becomes more immediate, the sensitivity to risk increases. Behaviors that were reinforced earlier in the process, such as engagement and exploration, are replaced by behaviors that reflect caution. Delays, requests for additional information, and the need for further validation are all expressions of this shift.

If the seller continues to operate as if the primary objective is to provide more information, the interaction can become misaligned.

The buyer is not seeking more detail.

They are seeking reassurance.

This is where the role of the seller changes again.

In earlier stages, the focus is on creating clarity around the problem and the solution. In the final stage, the focus shifts to reducing the perceived risk of acting on that solution.

This requires a different set of behaviors.

It involves acknowledging the concerns that may not have been expressed directly. It involves framing the decision in a way that makes the trade-offs clear. It involves helping the buyer see not only the benefits of moving forward, but the implications of not doing so.

These behaviors are not about pressure.

They are about alignment.

Romeo began to approach this stage differently.

Instead of reinforcing what had already been established, he focused on what remained uncertain. He asked questions that brought those uncertainties into the conversation, even when they were not explicitly raised.

"What would need to be true for this to feel like the right decision?"

"Where do you see the most risk in moving forward?"

"What concerns would someone else on your team raise if they were in this conversation?"

These questions did not always produce immediate answers.

But they created space for the buyer to articulate what had been implicit.

Once those concerns were visible, the conversation changed.

It became possible to address them directly, not with generic reassurance, but with specific examples, clear expectations, and a shared understanding of how the transition would be managed. The focus shifted from selling the solution to supporting the decision.

This is a subtle distinction, but an important one.

The buyer does not need to be convinced that the solution works.

They need to feel confident that it will work in their environment.

Fiona observed a similar pattern across her team.

The sellers who struggled to close deals often had strong engagement in the early and middle stages, but lost momentum as the decision approached. Their focus remained on reinforcing value, even as the buyer's focus shifted to risk.

The sellers who closed more consistently adjusted their behavior.

They recognized the change in the buyer's state and responded accordingly. They addressed concerns before they became obstacles. They framed the decision in a way that made the path forward clearer and the potential consequences of inaction more visible.

This alignment gradually reduced the frequency and duration of delays.

The connection to productivity at this stage is often overlooked.

When risk is not addressed, the decision process extends. Additional conversations are required, more stakeholders become involved, and the effort required to reach a conclusion increases. The deal consumes more time and attention, even as the likelihood of closing decreases.

When risk is addressed effectively, the decision becomes more efficient.

The buyer moves forward with greater confidence, and the need for additional validation is reduced. The process does not necessarily become faster, but it becomes more direct.

In this way, the consistency of behavior in the final stage has a disproportionate impact on overall productivity.

Technology can provide support here, but its role remains consistent with earlier stages.

AI can identify where deals tend to stall, highlight patterns in delayed decisions, and surface common concerns based on similar opportunities. This information can inform the seller's approach, helping them anticipate where risk may arise.

But it does not replace the interaction.

It does not ask the question or navigate the response. That remains the responsibility of the seller.

Looking back, Romeo understood that the final stage of the deal was not about closing in the traditional sense.

It was about creating the conditions in which the buyer could make a decision with confidence.

That required attention to what had not yet been resolved, and a willingness to engage with concerns that were not always visible.

It required behavior that was consistent with the buyer's need for clarity, not just the seller's desire for progress.

Behavioral Recap: What to Practice This Week

This week, examine how you are showing up in the final stages of your active deals. Are you creating confidence, or are you creating urgency? Those produce very different outcomes.

- For any deal sitting at proposal stage, schedule a live walk-through rather than sending a PDF. Walk through it together. Pause often. Invite edits and questions in real time.

- When a buyer hesitates or goes quiet, name it directly before you defend your position. "It sounds like something still feels unclear. What is it?" opens more than any amount of reinforcing value would.

- Match your pace to the buyer's confidence level. If they are moving slowly, the answer is not to move faster. The answer is to find out what is holding back their readiness.

- End every closing conversation with explicit clarity: what was decided, who owns what, and what the next concrete step is. Vague next steps at this stage are expensive.

- Follow up within 24 hours of any closing conversation with reassurance, not pressure. Confirm what was agreed, acknowledge any open items, and make the path forward feel simple.

Quick Self-Check: *How consistently did my closing behavior this week increase buyer confidence rather than buyer urgency?*

Next Chapter: The Power of Relationships

Chapter 11 explores what happens next, the part most sellers underinvest in and then wonder why renewals feel hard. You will learn how to reinforce trust immediately after the signature, how to create renewal readiness from day one, and how to expand accounts through relevance rather than pressure.

In the hybrid era, the easiest deals to close are the ones you have already earned the right to keep. Chapter 11 shows you how to earn that right, and how to build on it.

Chapter 11:
The Power of Relationships

Retention and Expansion in a Hybrid World

The moment a deal closes is often treated as an ending. Targets are hit, forecasts are updated, and attention shifts to what comes next. In most sales environments, the close is the point at which the work is considered complete, or at least complete enough to move on. The responsibility transitions, the urgency fades, and the interaction that once carried so much weight becomes one of many.

From the buyer's perspective, that moment feels very different.

Nothing has ended.

In many ways, it has just begun.

Romeo did not fully understand this until a deal he had been proud to win began to unravel.

The process leading up to the decision had been strong. The conversations were consistent, the concerns had been addressed, and the buyer moved forward with a level of confidence that suggested alignment. By the time the agreement was finalized, there was a shared expectation that the solution would deliver the value that had been discussed.

For a period of time, everything appeared to be on track.

Then the questions started.

They were not framed as objections. They were framed as clarifications. Requests for additional detail. Concerns about how certain aspects of the solution would be implemented. None of it suggested dissatisfaction, but the tone had changed. The confidence that had been present at the point of decision was no longer as stable.

As questions became more frequent, what had once been a clear path forward began to feel uncertain.

What Romeo came to understand was that the close had not resolved the buyer's concerns.

It had simply shifted them.

Before the decision, the focus had been on whether the solution was the right choice. After the decision, the focus moved to whether that choice would produce the expected outcome. The nature of the risk changed, but it did not disappear.

From a behavioral perspective, this shift is predictable.

When a decision is made, the immediate reinforcement comes from the act of choosing. There is a sense of progress, of resolution. But as the implementation begins, the consequences of that decision become more tangible. New variables are introduced, and the potential for friction increases.

If the behaviors that supported the decision are not carried forward, the initial confidence can erode.

This is where many sales processes create a gap.

The behaviors that were applied consistently leading up to the decision are often not maintained afterward. The attention to detail, the responsiveness, and the alignment that characterized the earlier interactions give way to a different set of priorities. The relationship becomes less active, and the continuity that had supported momentum begins to weaken.

For the buyer, this creates a disconnect.

The experience they had during the decision process is not fully reflected in the experience that follows. What had felt like a supported transition begins to feel more uncertain. Even when the solution itself is effective, the lack of continuity can introduce doubt.

Fiona began to see this pattern across accounts.

The deals that produced long-term value were not simply the ones that closed successfully. They were the ones where the behaviors that led to the decision were sustained after it. The same level of attention that had been applied during the evaluation was present during the early stages of implementation. Questions were addressed quickly. Context was maintained. The seller remained engaged in a way that reinforced the buyer's confidence.

In those cases, the relationship strengthened.

In others, the pattern was different.

The close marked a shift in behavior. Communication became less frequent, responses were less immediate, and the continuity that had supported the decision was disrupted. This compounded into friction. The buyer was required to navigate the implementation with less support, and the experience became more variable.

The outcome was not always failure.

But it was less stable.

This is where Behavioral Consistency extends beyond the sale itself.

It is not limited to the actions that move a deal forward. It includes the actions that sustain the value of that deal over time. The same principles apply. Behavior that is consistent, aligned, and responsive reinforces engagement. Behavior that is inconsistent introduces variability.

From a productivity standpoint, this has long-term implications.

When value is sustained, the effort required to maintain the relationship decreases. The buyer remains engaged, the likelihood of renewal increases, and opportunities for expansion emerge naturally. The relationship becomes more efficient, not because less is being done, but because the behaviors that support it are stable.

When value is not sustained, the opposite occurs.

Additional effort is required to address concerns, rebuild confidence, and reestablish alignment. The relationship consumes more time and attention, even as its potential diminishes.

Romeo began to adjust his approach after recognizing this pattern.

He remained engaged beyond the close, not in a way that was intrusive, but in a way that maintained continuity. He followed up on the initial stages of implementation, checked in on the areas that had been identified as potential risks, and ensured that the context of the decision was not lost.

These actions did not require a significant increase in effort.

They required consistency.

From a behavioral perspective, this can be understood as extending reinforcement.

The behaviors that created confidence during the decision are reinforced by similar behaviors during implementation. This strengthens the association between the seller and the positive outcome. That association compounds into future behavior, including continued engagement and willingness to expand the relationship.

Technology can support this process by providing visibility into usage, engagement, and performance. It can highlight where the solution is being adopted effectively and where it is not. This information can guide the seller's actions, helping them focus on the areas that require attention.

But, as in previous stages, technology does not replace the need for behavior.

It does not maintain the relationship.

That remains a function of how consistently the seller engages.

Looking back, Romeo realized that the close was not the conclusion he had once believed it to be.

It was a transition.

A shift from one set of behaviors to another, both of which required consistency to produce the desired outcome.

The difference was that the second set was often less visible, and therefore easier to neglect.

Once he recognized that, the way he approached the relationship changed.

Not dramatically, but consistently.

Behavioral Recap: What to Practice This Week

This week, invest thirty minutes in your existing accounts. Not in renewal conversations. In the proactive behaviors that make renewal conversations easy when they do come.

- Schedule at least one proactive check-in with a key account that has nothing to do with renewal timing or an upsell opportunity. Make it about them.

- Review usage or engagement data for your top accounts. Where is adoption strong? Where has it softened? Proactively reach out to the ones showing early signs of friction.

- Bring one specific insight to every customer touchpoint this week. Not a product update.

Something that helps them do their job better or think about their challenges differently.

- If you have an expansion conversation to have, make sure visible value has been delivered first. Sequence the ask after the win, not before it.

- Identify your top three happiest customers and reach out to each of them this week. Ask if they would be willing to share their experience with a peer or participate in a reference conversation. Do it now, while momentum is high.

Quick Self-Check: *How intentional was I this week about protecting and growing my existing relationships, not just managing them?*

Next Chapter: Sales Leadership and Team Growth in the Hybrid Age

Individual excellence creates momentum. Leadership determines whether it scales.

Chapter 12 zooms out from the individual seller to the team and the systems that either reinforce or undermine consistent behavior at scale. You will learn why coaching behaviors matters more than managing outcomes, how to protect the conditions your sellers need to do their best work, and how to build a culture where the right behaviors are not just trained but actually practiced, day after day, in the real work of selling.

Sustainable growth in the hybrid era does not come from a few exceptional sellers. It comes from leaders who know how to multiply the right behaviors, deliberately and on purpose.

Chapter 12:
Sales Leadership and Team Growth in the Hybrid Age

Why Coaching Behaviors Matters More Than Managing Numbers in Building Consistency at Scale

By the time Fiona stepped back to look at the team as a whole, the pattern she had been observing at the individual level was impossible to ignore.

Some sellers were improving in a way that felt stable. Their deals were not all wins, but the variability in their performance had narrowed. Conversations progressed with fewer resets, follow-ups maintained momentum, and decisions, when they came, felt more grounded. There was a sense that their approach held together across different situations.

Others showed flashes of the same capability, but not with the same reliability. A strong interaction would be followed by a weaker one. Momentum would build, then dissipate. The difference was not in effort or intent. It was in the consistency of behavior across time.

At that point, the question began to change again.

It was no longer how an individual seller could improve.

It was how that improvement could be made repeatable across a group.

Most sales organizations attempt to solve this problem through training.

They define what good looks like, deliver that definition in a structured format, and expect that it will translate into behavior. In some cases, it does. More often, it produces awareness without sustained change. Sellers leave with a clearer understanding of what they should be doing, but the application of that understanding remains uneven.

From a behavioral standpoint, this outcome is not surprising.

Knowledge does not produce consistency on its own. Behavior changes when it is observed, reinforced, and repeated within the environment in which it occurs. When those conditions are not present, the behavior is unlikely to stabilize, regardless of how well it is understood.

Fiona began to approach this differently.

Instead of focusing on broad concepts, she narrowed her attention to a smaller set of observable behaviors that appeared consistently in the deals that moved forward. She did not attempt to define every aspect of the process. She focused on the points where behavior had the most direct impact on outcomes.

Follow-up timing was one of those points. The consistency with which sellers responded after a conversation influenced whether momentum was maintained. Discovery depth was another. The ability to move beyond surface-level discussion and clarify the underlying issue shaped how decisions were made. The way next steps were framed and confirmed affected whether the process had direction.

These behaviors were not new.

What was different was the way they were treated.

They became the focus of observation.

Instead of asking whether a call was successful, Fiona asked whether these behaviors were present. Not in general terms, but in specific, observable ways. Did the follow-up occur within the expected window? Was the next step clearly stated and confirmed? Did the conversation reach the level of depth required to understand the buyer's situation?

This level of specificity changed the conversation.

It made performance visible in a way that could be acted on.

Once behavior was visible, it could be reinforced.

This did not take the form of broad recognition or generic feedback. It was tied directly to the behavior itself. When a seller executed a behavior effectively, it was acknowledged in that context. When a behavior was absent, the feedback focused on what needed to change in the next interaction.

In behavioral terms, this created a reinforcement loop.

The behavior was identified, the outcome was observed, and the continuity between the two was made explicit. This reinforced the likelihood that the behavior would be repeated.

Measurement followed naturally from this process.

Not as a way to evaluate individuals, but as a way to understand patterns across the team. By tracking the presence and frequency of key behaviors, Fiona could see where consistency was improving and where it was not. This information did not replace judgment, but it provided a clearer foundation for it.

It also revealed something that had been difficult to see before.

The variability in outcomes was closely tied to the variability in behavior.

When behavior stabilized, outcomes followed.

This is where the role of AI becomes more integrated.

At the individual level, it provides visibility into patterns that might otherwise go unnoticed. At the team level, it allows those patterns to be aggregated and analyzed in a way that highlights trends. It can show which behaviors are present in successful deals, how consistently they are applied, and where gaps exist.

Used in this way, it supports the development of consistency.

It does not replace the need for coaching.

It enhances it.

Fiona's role began to shift as well.

She spent less time reviewing outcomes and more time examining the behaviors that produced them. Her conversations with sellers became more focused. Instead of discussing what had happened in a deal, she worked through how it had unfolded. Where did the interaction align with the buyer's context? Where did it diverge? What behavior would be adjusted in the next conversation?

This level of focus required discipline.

It also produced clarity.

The impact of this approach became more increasingly visible.

The team's performance did not change dramatically from one period to the next. There were no sudden spikes that could be attributed to a single initiative. What changed was the pattern.

Deals moved with greater consistency. The variation between strong and weak performance narrowed. The effort required to maintain the pipeline decreased, even as outcomes became more stable.

This is the effect of Behavioral Consistency at scale.

From a productivity standpoint, the implications are significant.

When behavior is consistent across a team, the system becomes more efficient. Less time is spent correcting errors, reestablishing context, or recovering lost momentum. More time is spent building on what has already been established. The overall output becomes more predictable, not because variability has been eliminated, but because it has been reduced.

When behavior is inconsistent, the opposite occurs.

The system absorbs the cost of that inconsistency. Additional effort is required to compensate for missed actions, unclear communication, and misaligned interactions. Productivity appears high, but much of that activity is reactive.

Looking back, Fiona understood that the goal was not to control every aspect of the sales process.

It was to create an environment in which the behaviors that mattered most were more likely to occur.

That environment was built through observation, reinforcement, and measurement.

Not as isolated activities, but as part of a system that operated continuously.

The result was not perfection.

It was stability.

And in a complex environment, stability is what allows performance to scale.

Behavioral Recap: What to Practice This Week

This week, lead through behavior rather than instruction. The behaviors your team sees you practice are more powerful than any direction you give.

- Protect Golden Hours. Block the 10 to noon and 2 to 4 windows for your team's selling activity. Move internal meetings to the edges of the day, not the heart of it.

- Coach one specific observable behavior per rep this week. Not a theme or a concept. One specific action. What it looks like when done well and what the next conversation should improve.

- Listen to one real call recording from a seller on your team. Not a dashboard summary. The actual conversation. Coach from what you hear, not what the CRM recorded.

- Recognize effort publicly and specifically. Not "great job this week" but "I noticed how you slowed down when the buyer hesitated on Tuesday. That showed real judgment."

- Audit your own calendar before you audit anyone else's. Are your own behaviors modeling what you are asking your team to build?

Quick Self-Check: *How consistently did my leadership behaviors this week reinforce focus, clarity, and trust, rather than urgency, volume, and activity?*

Next Chapter: The Future of Hybrid Selling

Leadership sets the standard. The future tests whether it holds.

Chapter 13 looks ahead, not to predict what is coming but to examine what will always be true regardless of how the tools, channels, and buyer behaviors continue to evolve. You will see why AI will raise the bar rather than replace the seller, how Behavioral Revenue becomes a lasting advantage in a world of increasing automation, and what it means to build the behaviors today that will still matter five years from now.

The next era of sales will not belong to the loudest, the fastest, or the most automated. It will belong to the most clear, the most consistent, and the most trusted. Chapter 13 defines what it takes to be that seller.

Chapter 13:
The Future of Hybrid Selling
Staying Ahead of the Curve

By the time the system began to hold, the question shifted one final time.

It was no longer whether the approach worked. That had become clear in the pattern of results. Conversations were more aligned, follow-through was more consistent, and the variability that had once been difficult to explain had begun to narrow. The work felt different, not because it had become easier, but because it had become clearer.

The question now was how to sustain it.

Not for a single deal or a single quarter, but consistently in an environment that would continue to change.

It is easy to assume that once a system is in place, the problem is solved.

In practice, the opposite is often true.

A system creates visibility, and visibility introduces a new level of responsibility. Once behavior can be observed, it becomes harder to ignore. Patterns that were once hidden become apparent, and the gap between what is understood and what is executed becomes more difficult to justify.

From a behavioral standpoint, this is where consistency is tested.

The initial adoption of a behavior is often supported by focus and intention. As attention shifts and new demands emerge, that support erodes. Without reinforcement, even effective behaviors can fade, replaced by those that are easier or more familiar.

This is not a failure of discipline.

It is a characteristic of behavior.

Romeo noticed this in himself.

There were periods where his execution felt precise. He responded to signals with clarity, maintained continuity across interactions, and adjusted his approach in a way that aligned with the buyer's context. During those periods, the work felt controlled. Not predictable in outcome, but stable in process.

There were other periods where that precision slipped.

Not dramatically, but enough to be felt. A follow-up would come later than it should have. A question would remain unasked. A moment that required attention would pass without adjustment. None of these instances stood out on their own, but together they began to affect the overall pattern.

What made the difference was not awareness.

It was reinforcement.

In behavioral science, consistency is not maintained through intention alone. It is maintained through systems that reinforce the behavior over time. These systems can take many forms, but they share a common function. They bring attention back to the behavior, connect it to outcomes, and create conditions that make repetition more likely.

In the context of selling, this means creating structures that support observation, feedback, and adjustment on an ongoing basis.

It means revisiting interactions not only to understand what happened, but to identify how behavior contributed to the outcome. It means using available tools, including AI, to surface patterns that might otherwise go unnoticed. And it means maintaining a level of focus on the micro-behaviors that matter, even as other priorities compete for attention.

Fiona approached this as an ongoing process rather than a fixed solution.

She did not assume that consistency, once achieved, would sustain itself. Instead, she built reinforcement into the rhythm of the team's work. Conversations about performance continued to focus on behavior. Feedback remained specific and tied to observable actions. Patterns were revisited, not as a retrospective exercise, but as a way of maintaining alignment.

In doing so, behavior became part of the conversation, not just the outcome.

This had a stabilizing effect.

The role of AI within this system also evolved.

In earlier stages, it provided visibility into patterns that were difficult to see. As those patterns became more familiar, its function shifted toward reinforcement. It served as a reminder, a prompt, and a source of feedback that kept behavior aligned with the standard that had been established.

This is where its value becomes more durable.

Not as a tool for generating output, but as a mechanism for maintaining consistency.

From a productivity standpoint, the long-term effect of this approach is cumulative.

When behavior is reinforced consistently, the effort required to maintain performance decreases. Actions become more efficient, decisions become clearer, and the overall system becomes more stable. The gains are not always immediate, but they compound over time.

When behavior is not reinforced, the system tends to revert.

Effort increases, variability returns, and the work becomes less predictable. The same issues that once appeared to be resolved begin to resurface, not because the solution was ineffective, but because it was not sustained.

Looking ahead, the environment in which selling takes place will continue to evolve.

New tools will be introduced. Additional layers of data will become available. The ways in which buyers engage will shift in response to those changes. Each of these developments will create new opportunities, but they will also introduce new complexity.

What will not change is the role of behavior.

Regardless of how the tools evolve, the outcome will continue to be shaped by what happens in the moments that matter. The ability to recognize those moments, respond with clarity, and maintain consistency across them will remain the defining factor.

This is what it means to operate as a Hybrid Seller.

It is not a matter of balancing technology and human interaction.

It is a matter of integrating them in a way that supports consistent behavior.

It is recognizing that insight, no matter how advanced, does not create performance on its own. It must be connected to action, and that action must be repeated with enough consistency to shape outcomes.

The work, then, is not to master a set of techniques or to adopt a particular framework.

It is to develop a way of operating that brings attention to behavior, aligns it with context, and reinforces it over time.

That work does not end. It evolves.

And so, too, does the Hybrid Seller.

Behavioral Recap: What to Practice This Week

This week, focus on sustaining behavior, not chasing new tactics. The future will keep changing. Your consistency is what carries forward.

- Reinforce one behavior that already works. Don't add complexity. Identify a behavior that drives results and double down on executing it with precision.

- Build a simple reinforcement loop. At the end of each day, review one interaction and ask: What did I do? What worked? What needs to adjust? Keep it tight and repeatable.

- Use AI as a prompt, not a decision-maker. Let it surface patterns or signals, then slow down and choose your response intentionally. Behavior stays yours.

- Catch the small slips early. A delayed follow-up. A missed question. A rushed response. These are not isolated moments. They are early signals. Correct them in real time.

- Revisit one recent deal or interaction. Not to analyze the outcome, but to identify the behaviors that shaped it. What would you repeat? What would you refine?

Quick Self-Check: *How consistently did I reinforce the behaviors that drive performance, even as new tools, signals, and demands competed for my attention?*

CLOSING REFLECTION

If there is a single idea that carries through everything in this book, it is this.

Performance is not determined by what you know, but by what you do consistently when it matters.

The gap between those two is where most sellers struggle. It is also where the work lives.

Not in more information or better tools, but in the ability to recognize the moment, respond with clarity, and repeat the right behavior over time.

That is what creates separation.

And that is what defines the way you operate moving forward.

You are **The Hybrid Seller**.

EPILOGUE
The Quiet Advantage

There is a version of success in sales that is easy to recognize.

It is visible in the numbers, reflected in the outcomes, and often associated with moments that stand out. A large deal, a strong quarter, a sequence of wins that suggests momentum.

There is another version that is less visible, but more durable.

It is found in the pattern that underlies those outcomes. In the consistency of behavior that produces them. In the way a seller approaches each interaction, not as an isolated event, but as part of a system that holds together over time.

This advantage does not rely on intensity.

It relies on consistency.

It is easy to overlook because it does not announce itself.

It is built through small actions, repeated in a way that becomes predictable. It is reinforced through attention to detail, through the willingness to adjust, and through the discipline of maintaining alignment even when it would be easier not to.

In doing so, it becomes the foundation for everything else.

Technology will continue to change the way selling is done.

It will provide new tools, new insights, and new ways of interacting. It will make certain tasks easier and others more complex. It will create opportunities that did not exist before and challenge assumptions that once felt stable.

What it will not do is replace the need for consistent behavior.

The advantage, then, is not found in having access to more information.

It is found in the ability to act on that information in a way that is aligned, deliberate, and repeatable.

That ability is not built overnight.

It is developed through practice, reinforced through observation, and sustained through attention.

In the end, the difference between those who struggle and those who perform consistently is rarely dramatic.

It is found in the details.

In the moments that are easy to overlook.

In the behaviors that are applied not occasionally, but repeatedly.

That is where the work is.

That is where the advantage lies.

And that is where The Hybrid Seller lives.

APPENDIX

This section brings structure to what you've already seen throughout the book.

The chapters focused on how modern selling actually unfolds. Here, those ideas are organized into a system you can reference, apply, and refine over time.

If performance feels inconsistent, start with the Insight–Execution Gap. If you're looking to improve execution, focus on Behavioral Consistency and the micro-behaviors that drive it. If you're looking to align your overall approach, use the H.Y.B.R.I.D. Framework as your guide.

Start with one behavior, one interaction, and build from there.

The Insight–Execution Gap™

Why sellers "know" what to do but don't consistently do it.

Most sales breakdowns occur not because sellers lack information, but because the behavior never matches the intention.

The Insight–Execution Gap™ shows four stages:

Insight: You gain clarity—through training, coaching, analytics, or experience.

Intention: You commit to change; motivation spikes.

Execution: Daily pressures intervene, inconsistency creeps in, and habits revert unless reinforced.

Reinforcement: Small wins, coaching, rewards, and environmental cues strengthen the new behavior until it becomes durable.

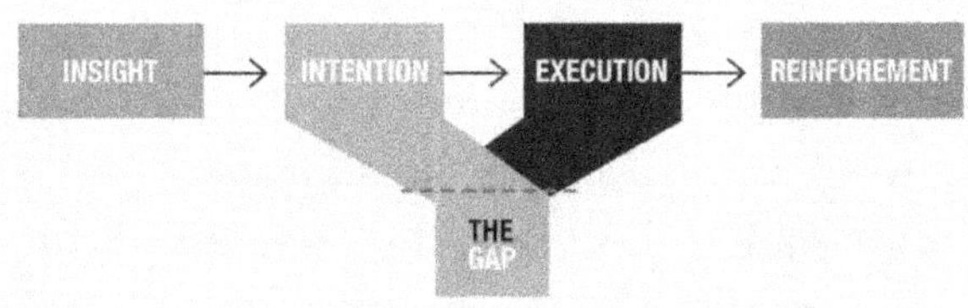

Figure Appendix A1 — The Insight–Execution Gap™

Behavior Consistency: The Engine of Execution

A behavioral operating system for modern selling.

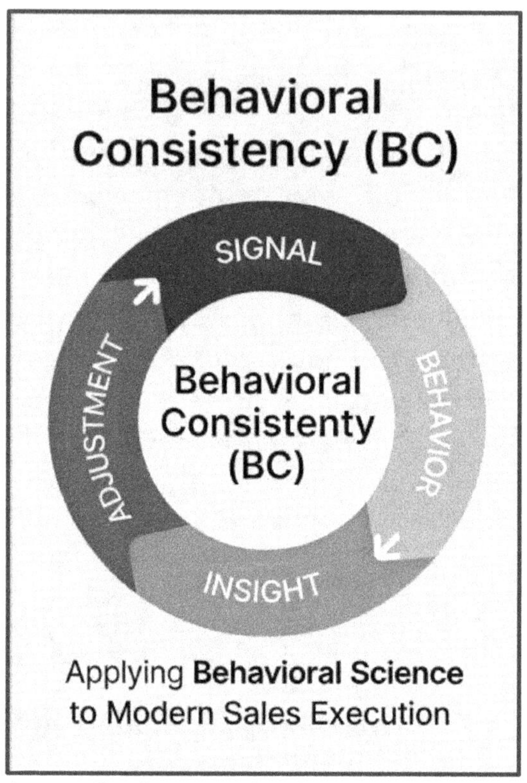

Figure Appendix A2 — Behavior Consistency Loop

Behavioral Consistency breaks down sales performance into four repeatable stages:

1. Signal: Buyer behavior, intent cues, data alerts, engagement spikes, or conversational clues that indicate opportunity or friction.

2. Behavior: The observable actions you take in response to those signals—pacing, questioning, follow-up quality, personalization, sequencing, and emotional presence.

3. Insight: The patterns revealed over time: which micro-behaviors correlate with progress, which stall or confuse buyers, and where gaps appear.

4. Adjustment: Small, intentional refinements that improve future performance: slowing down, asking one more layer, personalizing deeper, or changing timing.

Micro-Behaviors: The 15 High-Impact Actions That Predict Sales Success

Behavioral Consistency identifies 15 core **micro-behaviors** that most strongly correlate with pipeline momentum, deal progression, and closed-won outcomes. These behaviors appear across five selling domains and are reinforced by foundational ABA principles such as reinforcement, shaping, and environmental design.

THE 15 MICRO-BEHAVIORS ACROSS THE HYBRID SALES CYCLE

PROSPECTING	Follow-Up Latency · Trigger Responsiveness · Sequencing Depth · Touch Diversity
DISCOVERY	Agenda-Setting Consistency · Discovery Question Depth · Need Summarization Accuracy
PERSONALIZATION & ENGAGEMENT	Personalization Quality · Social Engagement Consistency · Content Alignment Behavior
HYBRID MEETING EXECUTION	Call Flow Adherence · Next-Step Commitment Setting
DEAL PROGRESSION & CLOSING	Stage-Advancing Behaviors · Follow-Up Quality · Multi-Threading Behavior

Figure A3 — The 15 Micro-Behaviors Across the Hybrid Sales Cycle

These aren't skills or traits, but rather **observable actions** that can be tracked, coached, and improved.

Prospecting Micro-Behaviors

1. **Follow-Up Latency** – How quickly you respond to buyer signals or inbound activity.
2. **Trigger Responsiveness** – Reacting promptly and meaningfully to intent data or buyer actions.
3. **Sequencing Depth** – Number and diversity of touchpoints used before abandoning a lead.
4. **Touch Diversity** – Use of multiple channels (email, phone, video, social) to build momentum.

Discovery Micro-Behaviors

5. **Agenda-Setting Consistency** – Opening meetings with structure, clarity, and expectations.
6. **Discovery Question Depth** – Asking layered questions that uncover context, impact, and root cause.
7. **Need Summarization Accuracy** – Reflecting buyer needs clearly before presenting solutions.

Personalization & Engagement Micro-Behaviors

8. **Personalization Quality** – Depth and relevance of tailored messaging.
9. **Social Engagement Consistency** – Frequency and value of interactions on digital platforms.
10. **Content Alignment Behavior** – Sharing relevant insights or resources aligned to buyer stage and role.

Hybrid Meeting Execution Micro-Behaviors

11. **Call Flow Adherence** – Staying aligned to a proven meeting structure while adapting to the buyer.
12. **Next-Step Commitment Setting** – Securing clear, mutual agreements at the end of every meeting.

Deal Progression & Closing Micro-Behaviors

13. **Stage-Advancing Behaviors** – Actions that move deals forward (diagnostic recaps, stakeholder alignment, etc.).
14. **Follow-Up Quality** – Clarity, personalization, and speed of follow-up messages.
15. **Multi-Threading Behavior** – Engaging multiple stakeholders to reduce single-point failure risk.

Behavioral Consistency Self-Assessment

Section 1: Prospecting Micro-Behaviors

Micro-Behavior	Description	Score (1–5)	Notes/ Examples
1. Follow-Up Latency	How quickly I respond to buyer activity or inbound signals	——	
2. Trigger Responsiveness	How well I act when intent data or alerts appear	——	
3. Sequencing Depth	Number + quality of coordinated touches before abandoning	——	
4. Touch Diversity	Use of email, phone, video, and social to build recognition	——	

Section 2: Discovery Micro-Behaviors

Micro-Behavior	Description	Score	Notes
5. Agenda-Setting Consistency	Whether I start meetings with clarity and expectations	——	
6. Discovery Question Depth	Ability to layer questions (surface → impact → root cause)	——	
7. Need Summarization Accuracy	How well I reflect buyer needs before proposing solutions	——	

Section 3: Personalization & Engagement Micro-Behaviors

Micro-Behavior	Description	Score	Notes
8. Personalization Quality	Depth of tailoring in messages, examples, and stories	——	
9. Social Engagement Consistency	How consistently I show up where buyers already are	——	
10. Content Alignment Behavior	Sharing insights aligned to stage, role, and priorities	——	

Section 4: Hybrid Meeting Execution Micro-Behaviors

Micro-Behavior	Description	Score	Notes
11. Call Flow Adherence	Staying aligned to a proven structure while adapting live	———	
12. Next-Step Commitment Setting	Securing clear mutual commitments in every meeting	———	

Section 5: Deal Progression & Closing Micro-Behaviors

Micro-Behavior	Description	Score	Notes
13. Stage-Advancing Behaviors	Actions that push opportunities forward intentionally	———	
14. Follow-Up Quality	Personalized, timely, value-adding follow-ups	———	
15. Multi-Threading Behavior	Engaging the full decision ecosystem early	———	

How Leaders Should Use This Scorecard

This scorecard is not a grading tool—it's a coaching accelerator. Leaders should use it weekly or bi-weekly to spot patterns, reinforce the right behaviors, and remove friction that slows sellers down. Score trends matter more than single numbers. Celebrate improvements, not just perfection. Your goal is to make the invisible visible: to turn instinct into insight, and insight into consistent execution. Used well, this becomes the engine of a high-performance, behavior-driven culture.

Behavioral Consistency Micro-Behavior Scorecard

A Practical Tool for Reps and Leaders to Measure What Matters

The following scorecard turns the 15 Behavioral Consistency Micro-Behaviors into a simple, repeatable system for evaluating performance, identifying patterns, and reinforcing the habits that drive real revenue outcomes.

Use it weekly or bi-weekly as part of coaching, pipeline reviews, or self-reflection.

Scoring Guide:

1 = Rarely

2 = Sometimes

3 = Often

4 = Consistently

5 = Always / Behaves like a top performer

Behavioral Insight Summary

Add your reflections here:

Which 3 micro-behaviors scored highest?

These represent your current strengths:

____

____

____

Which 2–3 scored lowest?
These are your leading indicators of future stalls:

____

____

What is one behavior you will adjust this week?
(Behavioral Consistency = small changes → big outcomes)

____

____

Which behaviors are missing from recorded calls or CRM notes?

____

____

Which behaviors correlate with the rep's recent wins?

____

____

Which behavior, if improved, would have the highest revenue impact?

____

____

Score Interpretation (Total Possible: 75)

Score Range	Meaning	Coaching Direction
60–75	Elite Hybrid Seller	Maintain strengths; refine precision
45–59	Strong Performer	Identify 2 micro-behaviors to improve
30–44	Inconsistent	Focus on structure, routine, and pacing
Below 30	At Risk	Reinforce basics; rebuild behavioral rhythm

The H.Y.B.R.I.D. Framework™: The Strategic Map for Modern Selling

The strategic backbone of modern selling. It provides the structure that organizes every behavioral move in the hybrid era.

H — Human Connection: Warmth, trust, presence, storytelling, emotional intelligence.

Y — Your Buyer's Journey: Understanding where the buyer really is—not where your process says they "should" be.

B — Bond Building: Momentum through consistency: personalized touchpoints, follow-up patterns, and recognition.

R — Relevance: Positioning, messaging, and examples tailored to the buyer's role, priority, and context.

I — Influence: Clarity, evidence, storytelling, value reframing, and objection handling rooted in psychology.

D — Drive Value Continuously: Post-sale excellence: renewals, expansions, referrals, and ongoing partnership.

How These Elements Work Together

Behavioral Consistency is the measurement and improvement engine.

The H.Y.B.R.I.D. Framework is the strategic map. **The Insight–Execution Gap** explains where performance breaks down.

Together, they create a system for translating insight into consistent execution, where performance becomes more predictable over time.

Framework Quick Reference Table

Framework	Purpose	What It Measures or Guides	Where Used in Book
Behavioral Consistency	Behavior measurement engine	Signals, micro-behaviors, insights, adjustments	Ch. 4
H.E.L.P. Model™	Ethical guardrails	Human-first interactions	Ch. 7
H.Y.B.R.I.D. Framework™	Strategic map	Connection → Value → Drive	Ch. 3
Insight–Execution Gap™	Behavior fall-off model	Where sellers stall	Ch. 4, 7

Figure Appendix A4 — Framework Quick Reference Table

QUICK REFERENCE PLAYBOOK

Chapter 1 – The Hybrid Modern Seller

The modern seller is not struggling because of a lack of tools or insight. They are overwhelmed by both. The shift is not from analog to digital. It is from **information to behavior.**

The Hybrid Seller recognizes that data does not create outcomes. Behavior does.

They use signals as guidance, not instruction. They interpret what is happening, choose the right action in the moment, and apply that action consistently. Variability decreases. Performance stabilizes.

The question is no longer, *What do I know?* It is, *What do I do next, and how consistently do I do it?*

Chapter 2 – The Buyer's Journey Has Changed

The buyer's journey is no longer linear. It is recursive, fragmented, and shaped long before the seller enters the conversation.

Buyers move between exploration, evaluation, and validation in loops. What appears to be progress can quickly become hesitation if clarity is not established.

The Hybrid Seller does not push buyers forward. They **help them make sense of where they are**.

Signals matter, but only when interpreted in context. Momentum is created not by speed, but by alignment.

Chapter 3 – The H.Y.B.R.I.D. Framework

The H.Y.B.R.I.D. Framework is not a script. It is a way of organizing attention.

Human Connection, Buyer's Journey awareness, Bond Building, Relevance, Influence, and Driving Value work together to guide behavior across interactions.

The goal is not to execute each element perfectly. It is to apply them **consistently and in alignment with the moment**.

When behavior aligns with context, conversations move. When it does not, friction appears.

Chapter 4 – Behavioral Consistency

Performance is not built on isolated moments of excellence. It is built on **repeatable patterns of behavior**.

The Behavioral Consistency Loop (Signal → Behavior → Outcome → Adjustment) explains how performance is formed and refined over time.

Most sellers do the right things occasionally. High performers do them consistently.

This is where productivity changes. Not through more activity, but through **more reliable execution of what matters**.

Chapter 5 – Consultative Selling in the Hybrid Era

Buyers no longer need information. They need clarity.

The role of the seller shifts from presenting to **helping the buyer think**. The most effective conversations are not the ones with the most information, but the ones that reduce uncertainty.

This requires different behaviors:

Listening beyond surface responses

Asking questions that reveal underlying concerns

Slowing down when clarity is needed

A single well-placed question can move a deal further than a full presentation.

Chapter 6 – Selling Across Channels

Selling no longer happens in a sequence. It happens across a system of interactions.

Email, calls, social, meetings, and content are not separate actions. They are part of one continuous conversation.

The Hybrid Seller focuses on **channel continuity**, ensuring that each interaction connects to the last and reinforces the overall message.

When continuity is strong, fewer interactions are needed. However, when it is weak, effort increases and momentum breaks.

Chapter 7 – AI, Automation, and the Discipline of Consistency

AI does not replace the seller. It exposes the seller.

It reveals patterns, highlights inconsistencies, and makes behavior visible. Used incorrectly, it increases activity without improving outcomes. Used correctly, it sharpens behavior.

The Hybrid Seller uses AI to:

Refine timing

Improve clarity

Identify missed signals

Not to do more, but to do **what matters with greater precision**.

Chapter 8 – The Moment That Matters

Deals do not break in obvious places. They break in subtle moments.

A hesitation. A pause. A partial answer.

These are the moments that determine whether a deal progresses or stalls.

The Hybrid Seller recognizes these signals and adjusts in real time. They ask one more question. They slow down. They create space for clarity.

Small behavioral adjustments in these moments have disproportionate impact on outcomes.

Chapter 9 – Follow-Through and Momentum

Momentum is not created in the meeting. It is preserved after it.

Follow-through is not administrative. It is behavioral.

Timing, clarity, and continuity determine whether the conversation continues or resets.

The Hybrid Seller:

Responds within the follow-through window

Reinforces what matters most

Makes next steps explicit

When follow-through is consistent, deals require less effort to move forward.

Chapter 10 – Decision and Risk

Deals rarely stall because of lack of value. They stall because of unresolved risk.

As decisions approach, buyers shift from evaluating benefits to avoiding mistakes.

The Hybrid Seller helps reduce risk by:

Surfacing unspoken concerns

Clarifying trade-offs

Supporting the decision process

Closing is not pressure. It is **removing uncertainty**.

Chapter 11 – After the Close

The close is not the end of the work. It is the beginning of the next phase.

Confidence does not carry forward automatically. It must be reinforced.

The behaviors that win the deal must continue after it:

Timely engagement

Clear communication

Ongoing alignment

When consistency continues, value compounds. When it stops, friction returns.

Chapter 12 – Building Consistency at Scale

Teams do not become consistent by accident. They become consistent by design.

Behavior must be:

Observed

Reinforced

Measured

Coached

The focus shifts from outcomes to the behaviors that produce them.

When behavior stabilizes across a team, performance becomes more predictable and more efficient.

Chapter 13 – The Work That Remains

There is no final version of the Hybrid Seller.

The environment will continue to change. Tools will evolve. Buyers will adapt.

What remains constant is the role of behavior.

The sellers who succeed are not the ones with the most information. They are the ones who **act on it consistently, when it matters most.**

Because the future doesn't belong to the loudest or most charismatic sellers, it belongs to the clearest, most consistent, and most trusted ones.

GLOSSARY

The Language of the Hybrid Seller™

As an educator, I learned early that every profession speaks its own language. Sales has shorthand. Education has acronyms. Special educators have an entirely different system of communication.

Hybrid selling is no different.

This field blends psychology, technology, data, storytelling, and human behavior. Without a shared language, it becomes difficult to apply consistently. The glossary that follows is not simply a set of definitions. It is a set of behavioral anchors.

If you remember nothing else from this book, remember this:

Performance is not a knowledge problem. It's a behavior problem.

When you understand the language, you begin to see the behavior.

When you see the behavior, you can change the outcome.

GLOSSARY OF KEY TERMS

A

Account Tiering

The process of segmenting accounts (e.g., Tier 1–3) based on opportunity size, engagement level, renewal value, or strategic priority.

Actionable Insight

A specific, behavior-generating takeaway that informs what a seller should do next, not just what they should know.

AI Acceleration

The increasing speed at which artificial intelligence enhances seller workflows, from drafting outreach to surfacing buyer intent.

B

Behavioral Consistency

The reliable execution of high-impact seller behaviors across time, channels, and buyer interactions.
Performance stabilizes when behavior stabilizes.

Behavioral Consistency Loop

A repeatable system that governs performance:
Signal → Behavior → Outcome → Adjustment
Used to observe, refine, and reinforce effective selling behaviors over time.

Behavioral Consistency

A behavioral operating system that measures, analyzes, and improves seller habits by connecting actions to outcomes.

Bond Building
Consistent actions that create familiarity, trust, and reliability
with a buyer through repeated, aligned interactions.

Buyer's Journey
The actual path a buyer takes from awareness to decision,
often non-linear and shaped before direct interaction with a
seller.

Buyer Signals
Observable cues that indicate buyer interest or friction,
including engagement, hesitation, timing, and communication
patterns.

C
Call Sequencing
A structured series of touchpoints (email, call, social, video)
aligned to buyer behavior and preferred communication
channels.

Channel Continuity
The consistency of experience across all communication
channels.
Buyers experience one conversation. Sellers often deliver
many disconnected ones.

Consultative Selling
A selling approach focused on helping buyers think clearly,
reduce uncertainty, and navigate decisions rather than simply
presenting solutions.

Credibility Markers
Proof elements (metrics, testimonials, case studies) that increase trust and reduce perceived risk.

D
Decision Confidence
The buyer's internal certainty that moving forward is the right choice, built through clarity and reduced risk.

Decision Friction
The resistance a buyer feels when moving toward a decision, often caused by uncertainty, risk, or lack of clarity.

Digital Fluency
The ability to move seamlessly across digital platforms while maintaining consistent presence and messaging.

Discovery Layering
A questioning technique that moves progressively deeper: surface → context → impact → root cause → future state.

Drive Value Continuously
The principle of delivering ongoing value beyond the close through engagement, outcomes, and partnership.

E
Engagement Signals
Measurable interactions that reveal buyer interest, hesitation, or confusion.

Execution Rhythm

A repeatable pattern of daily and weekly behaviors that drive consistent results.

Expansion Revenue

Revenue generated from existing customers through upsells, cross-sells, or added services.

F

Follow-Up Latency

The time it takes a seller to respond after a buyer signal. A critical micro-behavior.

Follow-Through (Behavioral Definition)

The continuation of the sales conversation after an interaction, where timing, clarity, and continuity either preserve or disrupt momentum.

Follow-Through Window

The critical period after a buyer interaction where momentum can be reinforced or lost.

Shorter latency strengthens continuity. Delays increase friction.

G

Golden Hours

Time blocks when buyers are most reachable and receptive. High performers protect these for selling activity.

H

Habit Formation Cycle
The process through which repeated actions become automatic behaviors, reinforced by cues and outcomes.

Hybrid Selling
A blended approach that integrates digital, personal, and social channels into one cohesive strategy.

H.Y.B.R.I.D. Framework™
The strategic architecture of the Hybrid Seller:
Human Connection → Your Buyer's Journey → Bond Building → Relevance → Influence → Drive Value Continuously

I

Influence
The ability to guide buyer thinking through clarity, reframing, and insight.

Insight–Execution Gap™
The gap between knowing what to do and consistently doing it, especially under pressure.
This gap is behavioral, not informational.

Intention Signal
A moment indicating a buyer's readiness to engage, deepen the conversation, or move forward.

L

Leading Indicators
Behaviors that predict future success, such as follow-up timing, discovery depth, and engagement quality.

Leverage Data Intelligently
Using data to inform decisions without allowing it to replace judgment.

M
Micro-Behaviors
Small, observable actions within interactions (timing, questioning, follow-up) that compound into major performance differences when applied consistently.

Momentum
The continuity of buyer engagement created through consistent, well-timed, and relevant seller behavior.

Moment That Matters
A subtle point within a conversation where buyer intent or hesitation is revealed.
These moments determine whether a deal progresses or stalls.

N
Noise (Sales Context)
Signals, data, or activity that do not meaningfully contribute to buyer progress or decision-making.

O
Objection Reframing
Responding to concerns in a way that clarifies thinking and reduces perceived risk.

Omnichannel Engagement
Coordinated selling across channels so the buyer experiences a unified interaction.

P

Pattern Recognition (Behavioral)

The ability to identify repeated behavioral sequences that lead to consistent outcomes.

Personalization at Scale

Tailoring communication so it feels individual, even when supported by automation.

Pipeline Velocity

The speed at which opportunities progress through stages, influenced by behavioral consistency.

Proof Library

A curated set of stories, metrics, and examples used to build trust quickly.

R

Relevance

The degree to which communication aligns with buyer context, priorities, and timing.

Retention Strategy

A system of behaviors designed to maintain customer engagement and long-term value.

Reinforcement Loop

A cycle of feedback and repetition that turns actions into stable behaviors.

Risk Compression

The process of reducing perceived buyer risk through clarity, alignment, and consistent behavior.

S
Sales Fluency
The ability to integrate product knowledge, buyer psychology, and behavioral precision into one cohesive approach.
Sequencing Depth
The extent to which a seller uses multiple channels to create engagement and momentum.

Signal Sensitivity
The ability to detect and interpret buyer signals accurately.

Signal vs. Noise Filtering
The ability to distinguish between meaningful signals and irrelevant activity.

Signal Interpretation
Assigning meaning to buyer behavior within context rather than reacting to data at face value.

T
Technological Agility
The ability to integrate AI and tools into workflows without losing judgment or human connection.

Timing Precision
The ability to act within the optimal moment to maintain relevance and momentum.

Trust Transfer
The process by which trust from one source influences another decision-maker.

U

Upsell Sequencing

Introducing expansion opportunities after value is demonstrated and risk is reduced.

Usage Intelligence

Tracking how customers use a product to inform retention and expansion strategies.

V

Value Framing

Presenting information in a way that clarifies outcomes rather than features.

Variability (Performance)

Fluctuation in outcomes caused by inconsistent behavior across interactions.

Virtual Presence

A seller's credibility, clarity, and professionalism expressed through digital channels.

AGENT INSIGHT™

At its core, The Hybrid Seller is built on foundational principles from behavioral science:

Signal → Behavior → Insight → Adjustment.

It's the same logic behind Applied Behavior Analysis: observe what's happening, identify what's working (and what isn't), make a targeted adjustment, and repeat until the outcome changes.

Most sellers don't lose deals because they lack information. They lose momentum in small, quiet ways: waiting too long to follow up, defaulting to safe language, skipping discovery, or letting the day dictate the plan. In a market moving this fast, the advantage isn't knowing more. It's executing consistently, especially when it's inconvenient.

That's why this companion app is included with the book. It is not "extra." It's the behavioral bridge between what you *believe* and what you *do*.

Agent Insight™ is your daily ritual. It helps you translate signals from your pipeline into one clear focus, one micro-mission, and one ready-to-use outreach script.

Access Agent Insight™ here:

You've read the science. Now make it real.

The edge is consistent execution of the right behaviors at the right time.

No cost. No upsell.

The only requirement is consistency.

The Hybrid Seller™ 10-Minute Daily Routine

The Hybrid Seller™ 10-Minute Daily Routine
You've read the science. Now apply it consistently—one small behavior at a time.

This routine turns insight into execution. No extra tools required. The only "cost" is consistency—show up for 10 minutes, make one decision, and take one clean action.

Do this once per day. Set a timer for 10 minutes.

TIME	STEP	WHAT TO DO
0:00–1:00	**Set your intention**	One sentence: "Today I will ______."
1:00–3:00	**Pipeline scan**	What changed since yesterday? What is the one deal that matters most right now?
3:00–5:00	**Name the constraint**	Choose the ONE blocker: Deal clarity, Champion strength, Economic case, Multi-threading, Timing, Competition, Decision process, Messaging fit.
5:00–8:00	**Create a micro-mission**	A 10–20 minute action you can complete today that reduces the constraint.
8:00–10:00	**Execute & log**	Send the message, book the meeting, make the call—then record what you did (1 line).

Today's 10-Minute Output (write it down)

1) My ONE constraint today is:

2) My micro-mission (10–20 min) is:

3) I will execute by (time):

Done (check one): ☐ Message sent ☐ Call made ☐ Meeting booked ☐ Follow-up logged

Optional: Use the Companion Tools when you want a ready-made execution card—but the routine works even without them.

SELECTED BIBLIOGRAPHY

The works listed below informed the thinking and applied frameworks presented in this book.

Ajzen, Icek. *Attitudes, Personality, and Behavior.* 2nd ed. Maidenhead, UK: Open University Press, 2005.

Ariely, Dan. *Predictably Irrational: The Hidden Forces That Shape Our Decisions.* New York: HarperCollins, 2008.

Bandura, Albert. *Social Learning Theory.* Englewood Cliffs, NJ: Prentice Hall, 1977.

Blount, Jeb. *Fanatical Prospecting: The Ultimate Guide to Opening Sales Conversations and Filling the Pipeline.* Hoboken, NJ: John Wiley & Sons, 2015.

Brynjolfsson, Erik, and Andrew McAfee. *The Second Machine Age: Work, Progress, and Prosperity in a Time of Brilliant Technologies.* New York: W. W. Norton & Company, 2014.

Cialdini, Robert B. *Influence: The Psychology of Persuasion.* Rev. ed. New York: Harper Business, 2021.

Covey, Stephen R. *The 7 Habits of Highly Effective People.* New York: Free Press, 1989.

Davenport, Thomas H., and Rajeev Ronanki. "Artificial Intelligence for the Real World." *Harvard Business Review* 96, no. 1 (January–February 2018): 108–116.

Dixon, Matthew, and Brent Adamson. *The Challenger Sale: Taking Control of the Customer Conversation*. New York: Portfolio/Penguin, 2011.

Dixon, Matthew, Ted McKenna, and Karen Freeman. *The JOLT Effect: How High Performers Overcome Customer Indecision*. New York: Portfolio, 2022.

Doerr, John. *Measure What Matters: How Google, Bono, and the Gates Foundation Rock the World with OKRs*. New York: Portfolio, 2018.

Duhigg, Charles. *The Power of Habit: Why We Do What We Do in Life and Business*. New York: Random House, 2012.

Ferrer, Manuel. *Behavioral Revenue: A Behavior-First Framework for Predictable Revenue Performance*. White paper. Fort Lauderdale, FL, 2024.

Fogg, B. J. *Tiny Habits: The Small Changes That Change Everything*. Boston: Houghton Mifflin Harcourt, 2019.

Heath, Chip, and Dan Heath. *Made to Stick: Why Some Ideas Survive and Others Die*. New York: Random House, 2007.

Kahneman, Daniel. *Thinking, Fast and Slow*. New York: Farrar, Straus and Giroux, 2011.

Kim, W. Chan, and Renée Mauborgne. *Blue Ocean Strategy.* Expanded ed. Boston: Harvard Business School Press, 2015.

McKinsey & Company. *The New B2B Growth Equation.* New York: McKinsey & Company, 2021.

McKinsey Global Institute. *The State of AI in 2023.* New York: McKinsey & Company, 2023.

Pink, Daniel H. *To Sell Is Human: The Surprising Truth About Moving Others.* New York: Riverhead Books, 2012.

Rackham, Neil. *SPIN Selling.* New York: McGraw-Hill, 1988.

Salesforce Research. *State of Sales.* San Francisco: Salesforce, annual reports.

Sinek, Simon. *Leaders Eat Last: Why Some Teams Pull Together and Others Don't.* New York: Portfolio, 2014.

Skinner, B. F. *Science and Human Behavior.* New York: Macmillan, 1953.

ACKNOWLEDGEMENTS

Writing *The Hybrid Seller*™ has been one of the most challenging, humbling, and ultimately transformative journeys of my career. This book is the product of my experiences, and the love, sacrifices, and encouragement of so many people who stood beside me when the path was uncertain and believed in me long before I fully believed in myself.

To my wife, Maria:

Your faith in me has never wavered. You've carried the weight of my long hours, travel schedules, and late-night writing sessions with grace and strength. Every step of this journey, every flight, every keynote, every chapter, was possible because of your love and partnership. Thank you for being my constant, my anchor, and my teammate in building a business, a legacy, and most importantly, a family.

To my daughters, Alyssa and Bryanna:

You are my *why*. You remind me every day why work should matter, why integrity counts, and why helping others is worth the effort. My hope is that this book gives you permission to dream boldly, take risks courageously, and lead with both heart and conviction.

To the memory of my grandparents:

You left everything behind in your native homeland so your children and grandchildren could have a future. I am that future. The sacrifices you made and the courage you carried live in these pages and in everything I do.

To my mother, Zoila; my sisters, Annette and Odalys; and my brothers, Jeovanee "Chef", and Lazaro "Laz" Valdes, whose memory stays with me.

Thank you for cheering me on and for always reminding me where I come from.

To the OGs: Juan "Bates" Parente, Willie "Coach" Chacon, Emilio "Flaco" Pacheco, and Jesus "JR" Rodriguez

You've always kept me grounded. No matter what's happening, you make sure I stay who I am. I'm grateful for the brotherhood and the way you continue to show up. I appreciate you more than you know.

To the sales leaders, colleagues, and mentors who shaped my journey:

From field ride-alongs to late-night debriefs, your lessons pushed me to grow, adapt, and rethink what it really means to serve buyers. The lessons I share here reflect the wisdom you modeled and the standards you held me to.

To the companies and teams I've worked with:

Thank you for letting me test, fail, adjust, and ultimately discover what it means to sell in a hybrid world.

To the readers:

Whether you're thumbing through these pages on a plane, reading late at night between calls, or stealing a few minutes during your lunch break, you are not alone. This book was written for you

ABOUT THE AUTHOR

Manuel "Dr. Manny" Ferrer, Ed.D., is a nationally recognized sales leader, an educator-turned-entrepreneur, and the creator of *The Hybrid Seller*™. His career began in education, where he was the recipient of the **Chappie James Most Promising Teacher Award** and **two-time Teacher of the Year** before transitioning into sales and quickly becoming a top performer and leader in the edtech industry.

In the field, Dr. Manny saw firsthand the collapse of the old sales playbook. Traditional inside and field selling were no longer enough, and the pandemic accelerated a permanent shift to hybrid models. He noticed a sharp divide: salespeople and companies that evolved thrived, while those clinging to the pre-2020 playbook instantly became outdated, hurting not just their sales teams, but their entire organizations.

Out of this experience, *The Hybrid Seller*™ was born: a blueprint that blends the timeless principles of trust and human connection with the latest tools, data, and AI-driven insights. Manny now equips sales professionals, leaders, and organizations with the mindset and skills to thrive in this new era.

When he's not writing, speaking, or coaching, Dr. Manny can be found traveling with his wife, Maria, and daughters, Alyssa and Bryanna, who have been his greatest supporters and inspiration throughout his journey.

THE NEXT CHAPTER

Your journey as a Hybrid Seller doesn't end with this book, it begins here.

If you'd like to access free resources, articles, and blogs that expand on the ideas in these pages, visit: **www.BreakingTheBell.com/TheNextChapter**.

You can also join **The Hybrid Seller Community**, where sellers, leaders, and educators-turned-entrepreneurs share insights, strategies, and support as we all continue to grow in this new era of sales.

Stay connected with me here:

LinkedIn: @drmanuelferrer

YouTube: @drmanuelferrer

X (Twitter): @drmanuelferrer

Email: DrManny@BreakingTheBell.com

Media & Guest Expert Appearances

As a recognized thought leader in educator career transitions and modern sales strategy, I'm available for:

Podcast and TV interviews

Guest articles and panel features

Expert commentary on workforce reform, education-to-entrepreneur pipelines, and minority-owned business development

For my speaker one-sheet, booking inquiries, or media requests, please email: **DrManny@BreakingTheBell.com**